A Pedagogical Introduction to Electroweak Baryogenesis

A Pedagogical Introduction to Electroweak Baryogenesis

Graham Albert White
ARC Centre of Excellence for Particle Physics at the Tera-scale, School of Physics and Astronomy, Monash University, Melbourne, Victoria 3800 Australia
and
Monash Centre for Astrophysics, School of Physics and Astronomy, Monash University, Melbourne, Victoria 3800 Australia

Morgan & Claypool Publishers

ISBN 978-1-6817-4457-5 (ebook)
ISBN 978-1-6817-4456-8 (print)
ISBN 978-1-6817-4459-9 (mobi)

DOI 10.1088/978-1-6817-4457-5

Version: 20161101

IOP Concise Physics
ISSN 2053-2571 (online)
ISSN 2054-7307 (print)

A Morgan & Claypool publication as part of IOP Concise Physics
Published by Morgan & Claypool Publishers, 40 Oak Drive, San Rafael, CA, 94903 USA

IOP Publishing, Temple Circus, Temple Way, Bristol BS1 6HG, UK

For Samuel White

Contents

Preface

We present a mostly self-contained pedagogical review of the theoretical background to electroweak baryogenesis as well as a brief summary of some of the other prevailing mechanisms for producing the asymmetry between matter and antimatter using the minimal supersymmetric Standard Model as a pedagogical tool whenever appropriate. This book covers an in-depth look at baryon number violation in the Standard Model, the necessary background in finite temperature field theory, plasma dynamics, and how to calculate the out-of-equilibrium evolution of particle number densities throughout a phase transition.

Acknowledgements

A lot of what I learnt about Baryogenesis was from Michael Ramsey Musolf and Csaba Balazs. I would also like to acknowledge the post-doctorates and students that answered many questions of mine especially Peter Winslow and Huaike Guo. I would also like to acknowledge David Paganin, Jonathan Kozaczuk, Giancarlo Pozzo, Kaori Fuyuto and Andrew Fowlie for useful comments regarding this work. Finally I would like to acknowledge my wife for enduring many nights of me working late during this project.

Author biography

Graham Albert White

Graham White grew up in north Queensland, Australia. He developed a love of physics after reading some popular science books as a teenager and decided he wanted to become a physicist shortly after. Graham White is a Doctoral Student at Monash University studying Baryogenesis and holds a masters degree from University of Kentucky and has studied Baryogenesis at the Amherst Center for fundamental interactions (University of Massachusetts Amherst) as a guest scholar. He currently lives in Melbourne with his wife and son.

A Pedagogical Introduction to Electroweak Baryogenesis

Graham Albert White

Chapter 1

Introduction

The origin of the baryon asymmetry of the Universe (BAU) is one of the deepest short-comings of our understanding of particle physics, as it cannot be explained within the Standard Model. For instance, one cannot simply set the baryon asymmetry as an initial condition as this would be washed out by inflation[1]. On the other hand, coinciding estimates of the baryon asymmetry using different techniques are a triumph of modern cosmology. The baryon asymmetry can be estimated by the deuterium abundance and from the cosmic microwave background (CMB), where the relative sizes of Doppler peaks in the temperature anisotropy are sensitive to the BAU. These two methods give the overlapping estimates of the baryon to entropy ratio[2] [2–4]

$$Y_B = \frac{n_B - \bar{n}_B}{s} \approx \frac{n_B}{s} = \begin{cases} (7.3 \pm 2.5) \times 10^{-11}, \text{ BBN} \\ (9.2 \pm 1.1) \times 10^{-11}, \text{ WMAP} \\ (8.59 \pm 0.11) \times 10^{-11}, \text{ Planck.} \end{cases} \tag{1.1}$$

This remarkable overlap in the estimates of the BAU from light element abundances (particularly deuterium) and baryon acoustic oscillations are shown in figure 1.1 and figure 1.2 respectively. Reproducing this estimate using particle physics makes one of the three pillars of the Standard Model of particle cosmology, the other two being inflation [5] and dark matter [6]. Like inflation and dark matter, it requires at least some additions to the Standard Model. Furthermore, like the other two pillars of the Standard Model of particle cosmology, there are a very large variety of models to explain this peculiar fact about our Universe. Two of the most elegant explanations

[1] For a recent attempted exception to this, albeit a fine-tuned one, see [1].

[2] Sometimes the baryon asymmetry is compared to the photon density, $n_B/n_\gamma \approx 7.04 Y_B$, rather than the entropy. However during the early Universe many particles are expected to be in thermal equilibrium making Y_B more convenient.

doi:10.1088/978-1-6817-4457-5ch1

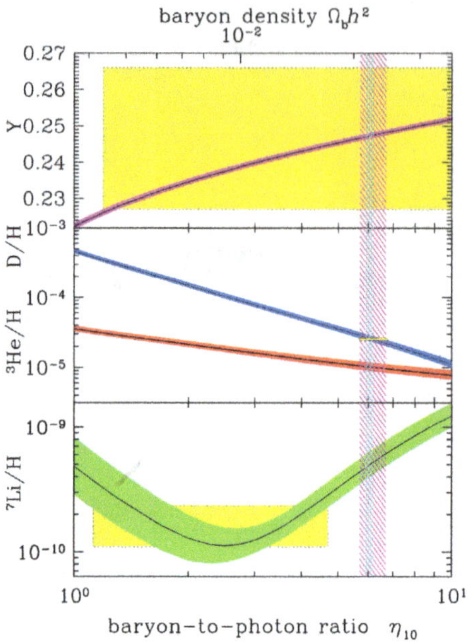

Figure 1.1. The abundances predicted by the Standard Model of Big Bang nucleosynthesis (BBN) [7] for 4He, D, 3He, and 7Li. Here the bands show the range for the 95% confidence level and the boxes indicate the light element abundances—the smaller boxes show 2.75σ statistical errors; the larger boxes 2.75σ statistical and systematic errors. The wide band indicates the BBN concordance range, whereas the vertical narrow band indicates the baryon asymmetry measured via the CMB given at the 95% confidence level. Reproduced from [2].

are leptogenesis [9] and the Affleck–Dine mechanism [10]. Unfortunately both tend to be well out of reach of the particle colliders of today and of the foreseeable future.

The focus of this review will be electroweak baryogenesis, which is the term for any mechanism that produces the matter–antimatter asymmetry during the electroweak phase transition. Such a scenario requires physics beyond the Standard Model that must couple relatively strongly to Standard Model particles and have masses that are not too far above the weak scale. Therefore, unlike Affleck–Dine baryogenesis or leptogenesis, electroweak baryogenesis has the tantalizing prospect of being tested, at least indirectly, by weak and TeV scale searches at the large hadron collider (LHC).

Apart from testability, electroweak baryogenesis has the attractive feature that coincides the breaking of the symmetry between particles and anti-particles with the spontaneous breaking of the one symmetry we know to be broken—electroweak symmetry. Unfortunately the literature on this exciting subject is somewhat opaque to newcomers. There are some very nice pedagogical introductions to small parts of the theoretical foundations of baryogenesis scattered throughout the literature if one digs hard enough[3]. However, the study of baryogenesis is arguably too decoupled

[3] For recent reviews see [11–14].

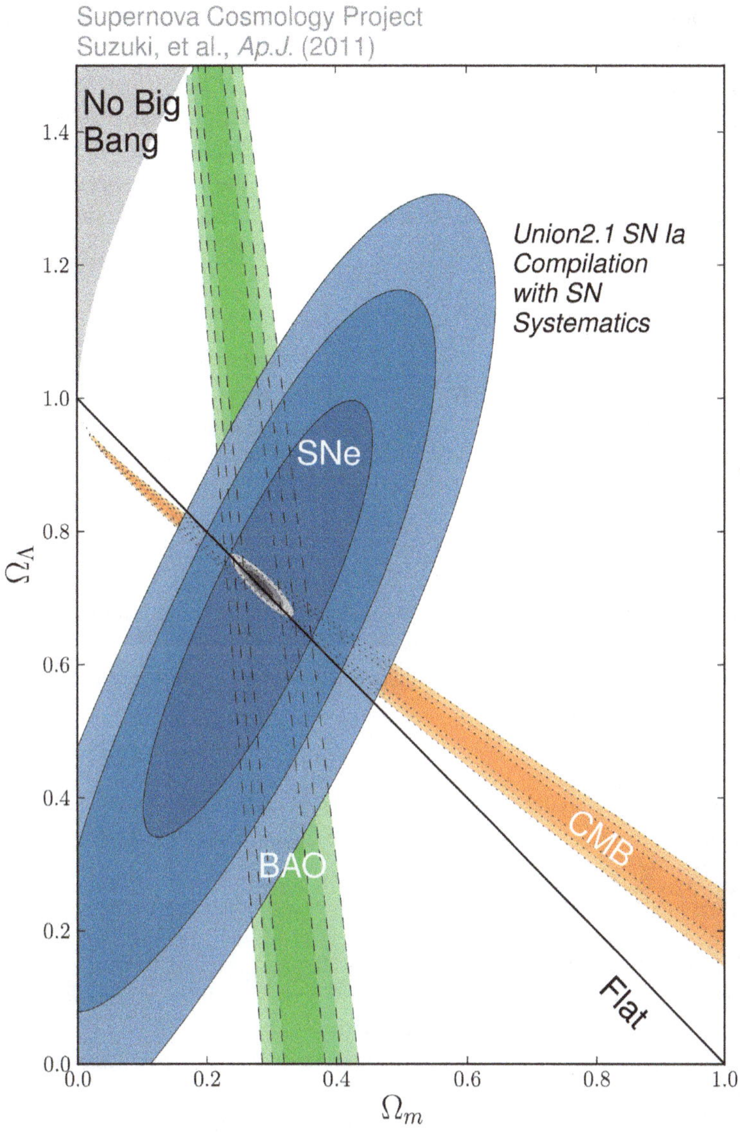

Figure 1.2. $\Omega_M - \Omega_\Lambda$ constraints due to CMB, baryon acoustic oscillations, and Supernova Cosmology Project Union2.1 SN constraints including SN systematic errors. Reproduced from [8].

from the rest of high energy physics given its promising phenomenological implications and the fundamental nature of the question it attempts to answer. Furthermore, the techniques one needs to learn to research in the field of electroweak baryogenesis have a large cross-over with other calculations in particle cosmology—including other models of producing the baryon asymmetry such as leptogenesis.

This primer will therefore give a mostly self-contained introduction to the field of *electroweak* baryogenesis, assuming the reader has a graduate level knowledge of particle physics, including dimensional regularization, some basic path integral techniques, the computation of amplitudes at tree and loop level, the Standard Model Lagrangian, and Big Bang cosmology, as well as a rudimentary knowledge of effective field theory and the minimal supersymmetric Standard Model (MSSM),[4] which we will use as a pedagogical tool where appropriate. There are of course many candidates for producing the BAU, but it is my hope that the theoretical foundations given in this book should make learning such mechanisms significantly more tractable.

References

[1] Krnjaic G 2016 Can the baryon asymmetry arise from initial conditions? arXiv: 1606.05344

[2] Eidelman S *et al* 2004 (Particle Data Group Collaboration) *Review of Particle Physics Phys. Lett.* B **592** 1

[3] Spergel D N *et al* 2003 (WMAP Collaboration) First year wilkinson microwave anisotropy Probe (WMAP) observations: determination of cosmological parameters *Astrophys. J. Suppl.* **148** A16

[4] Ade P A R *et al* 2014 Planck 2013 results. XVI. cosmological parameters *Astron. Astrophys.* **571** A16

[5] Bassett B A, Tsujikawa S and Wands D 2006 Inflation dynamics and reheating *Rev. Mod. Phys.* **78** 2

[6] Feng J L 2010 Dark matter candidates from particle physics and methods of detection *Annu. Rev. Astron. Astrophys.* **48** 495–545

[7] Cyburt R H *et al* 2008 An update on the big bang nucleosynthesis prediction for 7Li: the problem worsens *J. Cosmol. Astropart. Phys.* JCAP11(2008)012

[8] Suzuki N *et al* 2012 The Hubble Space Telescope Cluster Supernova Survey. V. Improving the dark-energy constraints above z > 1 and building an early-type-hosted supernova sample **756** 85

[9] Fukugita M and Yanagida T 1986 Baryogenesis without grand unification *Phys. Lett.* B **174** 1

[10] Affleck I and Dine M 1985 A new mechanism for baryogenesis *Nucl. Phys.* B **249** 361

[11] Morrissey D E and Ramsey-Musolf M J 2010 Electroweak baryogenesis *New J. Phys.* **14** 12

[12] Cline J M 2012 *Baryogenesis* arXiv: 0609145(hep-ph)

[13] Riotto A and Trodden M 1999 Recent progress in baryogenesis *Annu. Rev. Nucl. Part. Sci.* **49** 46

[14] Trodden M 1999 Electroweak baryogenesis *Rev. Mod. Phys.* **71** 5

[15] Kuroda M 1999 Complete lagrangian of MSSM arXiv: 9902340(hep-ph)

[16] Csaki C 1996 The minimal supersymmetric standard model (MSSM) *Mod. Phys. Lett.* A **11.08** 599–613

[4] If you lack knowledge of the MSSM see [15,16] for an introduction.

Chapter 2

The Sakharov conditions

Perhaps the most convincing motivation for requiring an explanation for the BAU is the fact that inflation, a cornerstone of modern cosmology, will wash out any initial BAU. Before inflation was proposed, however, Sakharov proposed that any explanation of the BAU must satisfy three conditions, now famously known as the 'Sakharov conditions'. These conditions are:

- Violation of baryon number conservation.
- Violation of C and CP.
- Departure from equilibrium.

The conditions are intuitively obvious as they basically say in crude terms that you need a process that allows you to create baryons, treat particles differently to anti-particles[1] and the future different to the past, and you need the process to change something in the Universe [1]. Even though the last condition is the most obvious it actually has a well-known loophole if CPT invariance is violated in a theory [2]. This can be seen from the proof of the third condition. Consider the equilibrium average of the baryon asymmetry B using standard thermal quantum mechanics[2],

[1] A mild subtlety is that C or CP violation separately does not achieve this. One requires both C and CP violation as a necessary condition for the rate of *total* baryon production to be different to total anti-baryon number.

[2] An astute reader might note that this argument implies that the baryon asymmetry is zero when the Universe returns to equilibrium. Indeed, processes that violate baryon number conservation do conspire to return the baryon number to zero on a time scale much larger than the age of the Universe. The key point of the argument however is that the baryon number of the Universe should be zero without some departure from equilibrium (or a violation of CPT).

doi:10.1088/978-1-6817-4457-5ch2

$$
\begin{aligned}
\langle B\rangle &= \mathrm{Tr}\left[\mathrm{e}^{-\beta\mathcal{H}}B\right]\\
&= \mathrm{Tr}\left[(CPT)(CPT)^{-1}\mathrm{e}^{-\beta\mathcal{H}}B\right]\\
&= \mathrm{Tr}\left[\mathrm{e}^{-\beta\mathcal{H}}(CPT)^{-1}B(CPT)\right]\\
&= -\mathrm{Tr}\left[\mathrm{e}^{-\beta\mathcal{H}}B\right].
\end{aligned}
\tag{2.1}
$$

The proof is quite elementary but does reveal that in asserting that baryon asymmetry production requires a departure from equilibrium you are implicitly assuming CPT invariance. This, however is a safe assumption in the vast majority of models including any this author is aware of that use the electroweak baryogenesis mechanism. Despite these conditions being intuitively obvious they are useful for the organization of this book, which is structured around these conditions. Remarkably one can, in principle, satisfy all three conditions using Standard Model particle content. However, the precise values of parameters within the Standard Model rule out it being sufficient to explain the BAU.

References

[1] Sakharov A D 1967 Violation of CP invariance, C asymmetry, and baryon asymmetry of the universe *J. Exper. Theor. Phys.* **5** 24–7
Sakharov A D 1991 Violation of CP invariance, C asymmetry, and baryon asymmetry of the universe *Sov. Phys. Usp.* **34** 392–3

[2] Bertolami O, Colladay D, Kostelecký V A and Potting R 1997 CPT violation and baryogenesis *Phys. Lett.* B **395** 178–83

Chapter 3

Baryon number violation in the Standard Model

The baryon number is a classically conserved quantity in the Standard Model as can be demonstrated by Noether's theorem or even a cursory look at the available interactions. This intuition breaks down when one quantizes the Standard Model, due to so-called anomalous processes, which will be described in this chapter. That some fundamental symmetries may not survive the process of quantization is highly counter-intuitive. This is not just a feature of field theory. It was shown in [1, 2] that even standard quantum mechanics with certain potentials—specifically a delta function or a $1/r^2$ potential—have anomalous violations of conservation laws. The key anomaly of interest is the so-called chiral or axial anomaly. This anomaly, combined with the fact that the SU(2) gauge symmetry is a symmetry of left-handed particles only, will conspire to fulfil Sakharov's baryon number violation condition.

3.1 The axial anomaly

Let us warm up, however, by considering the simplest possible fermionic field theory, a single massless, free fermion field with a U(1) gauge symmetry. Such a theory has a global phase symmetry. Less obvious is that such a model's Lagrangian is also invariant under the following transformation

$$\psi \to e^{i\theta\gamma_5}. \tag{3.1}$$

From these symmetries we can of course use Neother's theorem to derive conserved currents. The Noether currents associated with these two symmetries are

$$J_\mu = \bar{\psi}\gamma_\mu\psi \tag{3.2}$$

$$J_{5\mu} = \bar{\psi}\gamma_5\gamma_\mu\psi, \tag{3.3}$$

doi:10.1088/978-1-6817-4457-5ch3

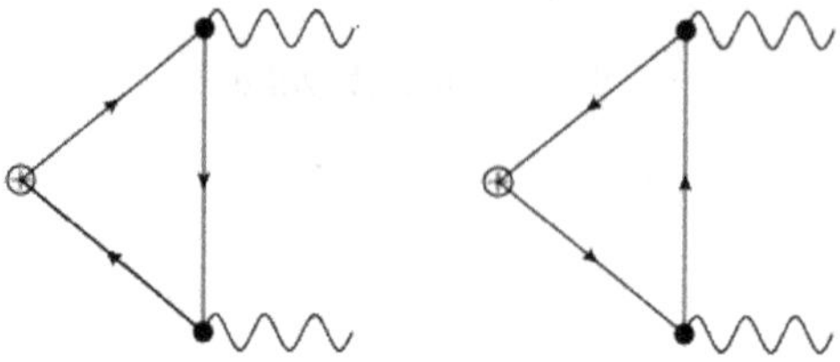

Figure 3.1. The famous triangle diagram responsible for anomalous quantum violations of classical chiral symmetry. The circle denotes a γ_5 on the vertex.

which implies Ward identities $\partial^\mu J_\mu = 0$ and $\partial^\mu J_{5\mu} = 0$. Obviously we expect the product of conserved currents to also be a conserved quantity. Therefore we should not even need to bother to calculate the divergence of the following quantity:

$$-\mathrm{i}\Gamma_{\mu\nu\lambda}(x_1, x_2, x_3) = \langle 0|T\big(J_{5\mu}(x_3)J_\nu(x_1)J_\lambda(x_2)\big)|0\rangle. \tag{3.4}$$

Amazingly this quantity is not zero. This remarkable result is known as the 'triangle anomaly' [3, 4] due to the fact that the Feynman diagram related to this amplitude has three vertices, as shown in figure 3.1. In the author's opinion the most effective way to learn the axial anomaly is through direct calculation. We present a review of what we believe is probably the simplest derivation in the literature using operator methods with dimensional regularization, followed by the Fujikawa method which is a path integral derivation.

3.1.1 Dimensional regularization

Let us calculate this example using the ever-familiar dimensional regularization. A potential ambiguity that can arise in such a calculation is that there is no clear way to generalize the matrix γ_5 to $4 - 2\epsilon$ dimensions although various proposals exist [5]. We follow the approach of [6], which demonstrated that no knowledge of any of γ_5 properties is needed to calculate the triangle anomaly[1]. The axial current to one-loop order due to the triangle anomaly is given by

$$\langle 0|J_5^\mu(x)|0\rangle = \int \frac{\mathrm{d}^D p}{(2\pi)^D}\frac{\mathrm{d}^D k}{(2\pi)^D}\mathrm{e}^{\mathrm{i}q\cdot x}A_\nu(p)A_\lambda(k)\mathcal{M}^{\mu\nu\lambda}, \quad q = k + p. \tag{3.5}$$

The above amplitude relates to the two diagrams in figure 3.1, and the second diagram is obtained from the first by merely interchanging (p, ν) and (k, λ). Hence, it suffices to compute one diagram

$$\mathcal{M}^{\mu\nu\lambda}_{(1)}(k, p) = -\mathrm{i}e^2\int \frac{\mathrm{d}^D l}{(2\pi)^D}\left\{\gamma^\mu\gamma_5\frac{\not{l} - \not{k}}{(l-k)^2 + \mathrm{i}\epsilon}\gamma^\lambda\frac{\not{l}}{l^2 + \mathrm{i}\epsilon}\gamma^\nu\frac{\not{l} + \not{p}}{(l+p)^2 + \mathrm{i}\epsilon}\right\}, \tag{3.6}$$

[1] Part of this calculation has some cross-over with the calculation in [7].

where the curly brackets denote a trace operation and the photon momenta are on-shell. Although we introduced this section stating that we are considering a massless fermion model, we will use a mass to regularize the integrals and take the limit $m^2 \to 0$ at the end. Let us integrate over loop momentum and contract with q_μ. The result we can write as the sum of two terms

$$\mathrm{i}q_\mu \mathcal{M}^{\mu\nu\lambda}_{(1)} = \mathcal{A}^{\nu\lambda}(k, p) + \mathcal{B}^{\nu\lambda}(k, p), \tag{3.7}$$

where

$$\begin{aligned} \mathcal{A}^{\nu\lambda}(k, p) = &\frac{\mathrm{i}e^2}{32\pi^2}\int_0^1 \mathrm{d}x \int_0^{1-x} \mathrm{d}y\left[\frac{1}{\bar{\epsilon}} - \ln\left(\frac{\Delta}{\mu^2}\right)\right] \\ &\times \left\{\not{q}\gamma_5\gamma^\alpha\gamma^\lambda\gamma_\alpha\gamma^\nu(x\not{k} + (1-y)\not{p}) - \not{q}\gamma_5(y\not{p} + (1-x)\not{k})\right. \\ &\times \left.\gamma^\lambda\gamma^\alpha\gamma^\nu\gamma_\alpha - \not{q}\gamma_5\gamma^\alpha\gamma^\lambda(y\not{p} - x\not{k})\gamma^\nu\gamma_\alpha\right\} \end{aligned} \tag{3.8}$$

and

$$\begin{aligned} \mathcal{B}^{\nu\lambda}(k, p) = &-\frac{\mathrm{i}e^2}{16\pi^2}\int_0^1 \mathrm{d}x \int_0^{1-x} \mathrm{d}y \\ &\times \frac{\left\{\not{q}\gamma_5((x-1)\not{k} - y\not{p})\gamma^\lambda(x\not{k} - y\not{p})\gamma^\nu(x\not{k} + (1-y)\not{p})\right\}}{\Delta}. \end{aligned} \tag{3.9}$$

In the above we have defined the parameters

$$\Delta = -q^2xy + m^2, \qquad \frac{1}{\bar{\epsilon}} = \frac{1}{\epsilon} - \gamma + \ln 4\pi. \tag{3.10}$$

The tensor $\mathcal{A}^{\nu\lambda}$ contains a divergent piece. If we employ the identities

$$\gamma^\alpha\gamma^\mu\gamma_\alpha = (-2 + 2\epsilon)\gamma^\mu, \qquad \gamma^\alpha\gamma^\mu\gamma^\lambda\gamma^\nu\gamma_\alpha = -2\gamma^\nu\gamma^\lambda\gamma^\mu + 2\epsilon\gamma^\mu\gamma^\lambda\gamma^\nu \tag{3.11}$$

along with the cyclic property of traces, we can reduce this tensor to the form

$$\begin{aligned} \mathcal{A}^{\nu\lambda}_{\text{div.}} = &\frac{\mathrm{i}e^2}{96\pi^2\bar{\epsilon}}\Big[4p^\lambda\{\gamma^\nu\not{q}\gamma_5\} - 4k^\nu\{\gamma^\lambda\not{q}\gamma_5\} - 2p^\nu\{\gamma^\lambda\not{q}\gamma_5\} + 2k^\lambda\{\gamma^\nu\not{q}\gamma_5\} \\ &+ 2g^{\nu\lambda}\{\not{k}\not{p}\gamma_5 - \not{p}\not{k}\gamma_5\}\Big]. \end{aligned} \tag{3.12}$$

So all traces with four Dirac matrices have already canceled and we are left with a series of terms that are zero in four dimensions. However, we need not make an assumption about the value of these traces. The term multiplying $1/\bar{\epsilon}$ is anti-symmetric under the interchange (p, ν) and (k, λ) so it cancels when the two diagrams are added, giving

$$\mathcal{A}^{\nu\lambda}_{\text{div.}}(k, p) + \mathcal{A}^{\lambda\nu}_{\text{div.}}(p, k) = 0. \tag{3.13}$$

The finite part of $\mathcal{A}^{\nu\lambda}$ comes from the $\mathcal{O}(\epsilon)$ part of equation (3.11) and the ln Δ part in equation (3.8). The result in the massless limit reads

$$\lim_{\epsilon\to 0}\left[\mathcal{A}^{\nu\lambda}(k, p) + \mathcal{A}^{\lambda\nu}(p, k)\right]_{m=0} = \frac{\mathrm{e}^2}{6\pi^2}\epsilon^{\lambda\nu\alpha\beta}p_\alpha k_\beta. \tag{3.14}$$

Note that the above tensor is infrared finite, so the inclusion of a mass regulator was a convenient calculation tool. Moving to $\mathcal{B}^{\nu\lambda}(k, p)$, we find that the potentially singular terms cancel identically so we can take the limit $D \to 4$ and write

$$\lim_{\epsilon\to 0}\left[\mathcal{B}^{\nu\lambda}(k, p) + \mathcal{B}^{\lambda\nu}(p, k)\right]_{m=0} = \frac{\mathrm{e}^2}{3\pi^2}\epsilon^{\lambda\nu\alpha\beta}p_\alpha k_\beta. \tag{3.15}$$

Putting everything together yields the well-known, but non-zero result

$$\langle 0|\partial \cdot J_5|0\rangle = \frac{\mathrm{e}^2}{2\pi^2}\epsilon^{\mu\nu\alpha\beta}\int \frac{\mathrm{d}^D p}{(2\pi)^D}\frac{\mathrm{d}^D k}{(2\pi)^D}\mathrm{e}^{\mathrm{i}q\cdot x}A_\nu(p)A_\lambda(k)\, p_\alpha k_\beta. \tag{3.16}$$

Thus we have have found by explicit calculation the anomalous violation of chiral current conservation. We have chosen a particular way of performing the calculation that makes it maximally clear that the anomaly is not some ill effect of how we regulate our integrals. Indeed the anomaly is experimentally demonstrated in predicting the correct rate for pion decay $\pi^0 \to \gamma\gamma$. Furthermore, as remarked in the previous section, it has been shown to happen in ordinary quantum mechanics and the above calculation has been repeated with multiple different regularization schemes. Nonetheless we present one more derivation in the next subsection, since it naturally organizes the anomalies in the Standard Model such that it is straightforward to show that lepton and baryon currents have anomalous currents.

3.1.2 The Fujikawa method

Since the anomaly is independent of the regularization scheme we will calculate the anomalies using the path integral method with a hard cutoff regularization scheme. The regularization free nature of the anomaly is more opaque in this scheme, so I will assume that the reader understood the results of the previous section where things were more clear. The cost of this opaqueness will be justified by the simplicity of the calculation that follows. Let us begin by returning to the anomalous chiral current

$$J^5_\mu = \bar{\psi}\gamma_5\gamma_\mu\psi. \tag{3.17}$$

As in the previous section, the triangle diagram anomalously violates the classical conservation law derived through Noether's theorem. This time our fermions interact with Standard Model gauge bosons. Path integral techniques are far more efficient and systematic than trying to find all anomalous operators and hoping you have not missed any. The path integral's approach begins with simply adding the current to the Lagrangian [8, 9]

$$W\left[a_\mu, A_\lambda^c\right] = \int [\mathrm{d}\psi][\mathrm{d}\bar{\psi}]\mathrm{e}^{\mathrm{i}\int \mathrm{d}^4x \mathcal{L}_{\mathrm{QCD}} - a_\mu J_5^\mu}. \tag{3.18}$$

We then vary W with respect to a_μ by an infinitesimal function $\partial_\mu \beta$ in order to test whether W is invariant under such a change. Using the definition of functional derivatives

$$\begin{aligned}
\delta[\ln W] &= \ln\left[W\left(a_\mu - \partial_\mu\beta, A_\mu^b\right)\right] - \ln\left[W\left(a_\mu, A_\mu^b\right)\right] \\
&= -\int \mathrm{d}^4x \frac{\delta \ln W}{\delta a^\mu} \\
&= \mathrm{i}\int \mathrm{d}^4x \bar{J}_5^\mu \partial_\mu\beta \\
\delta[\ln W] &= -\mathrm{i}\int \mathrm{d}^4x \partial_\mu \bar{J}_5^\mu \beta(x).
\end{aligned} \tag{3.19}$$

So if $\delta[\ln W] = 0$, $\partial_\mu J_5^\mu = 0$ and the chiral current is conserved. To test for an anomaly one determines if the statement

$$W\left[a_\mu - \partial_\mu\beta, A_\lambda^c\right] = W\left[a_\mu, A_\lambda^c\right] \tag{3.20}$$

holds true. Let us begin by seeing if there is a change of variables that takes our primed Lagrangian back to the original form. That is

$$L'_{\mathrm{QCD}}\left(\psi', \bar{\psi}', A_\mu^a\right) \equiv L_{\mathrm{QCD}}\left(\psi', \bar{\psi}', A_\mu^a\right) - \left(\partial_\mu\beta\right)J_5^\mu \to L_{\mathrm{QCD}}\left(\psi, \bar{\psi}, A_\mu^a\right). \tag{3.21}$$

The appropriate set of transformations is

$$\psi' = \left(1 - \mathrm{i}\beta\gamma_5\right)\psi \sim \mathrm{e}^{-\mathrm{i}\beta\gamma_5}\psi \tag{3.22}$$

$$\bar{\psi}' = \bar{\psi}\left(1 - \mathrm{i}\beta\gamma_5\right) \sim \bar{\psi}\mathrm{e}^{-\mathrm{i}\beta\gamma_5}. \tag{3.23}$$

This can be verified by direct substitution in which one ignores all terms $\mathcal{O}(\beta^2)$ and uses the anti-commutivity of the gamma matrices. The only other modification to make to our function W is the measure

$$[\mathrm{d}\psi][\mathrm{d}\bar{\psi}] \to [\mathrm{d}\psi'][\mathrm{d}\bar{\psi}'] \tag{3.24}$$

in which we will pick up a Jacobian determinant $\mathcal{J}$. We will later verify that $\mathcal{J}$ is independent of ψ and $\bar{\psi}$ so we can take it out of the integral

$$\begin{aligned}
W\left[a_\mu - \partial_\mu\beta, A_\mu^a\right] &= \int [\mathrm{d}\psi'][\mathrm{d}\bar{\psi}']J\mathrm{e}^{\mathrm{i}\int \mathrm{d}^4x L_{\mathrm{QCD}} - a_\mu J_5^\mu} \\
&= J\int [\mathrm{d}\psi'][\mathrm{d}\bar{\psi}']\mathrm{e}^{\mathrm{i}\int \mathrm{d}^4x L_{\mathrm{QCD}} - a_\mu J_5^\mu} \\
&= JW\left[a_\mu, A_\mu^a\right].
\end{aligned} \tag{3.25}$$

Therefore we can write the variation of ln W exclusively in terms of the Jacobian

$$\begin{aligned}\delta[\ln W] &= \ln\left[W\left(a_\mu - \partial_\mu\beta,\, A_\mu^a\right)\right] - \ln\left[W\left(a_\mu,\, A_\mu^a\right)\right] \\ &= \ln\left[\mathcal{J}W\left(a_\mu,\, A_\mu^a\right)\right] - \ln\left[W\left(a_\mu,\, A_\mu^a\right)\right] \\ &= \ln \mathcal{J}.\end{aligned} \tag{3.26}$$

Our Jacobian is of Grassman valued functions so it is given by the inverse of the functional determinant of our transformations

$$\mathcal{J} = \mathrm{Det}\,[\mathrm{e}^{\mathrm{i}\beta\gamma_5}\,\mathrm{e}^{\mathrm{i}\beta\gamma_5}]^{-1}. \tag{3.27}$$

We can use a simple identity, Det $C = \exp[\mathrm{Tr}\ln C]$, to write this determinant in terms of a trace

$$\mathcal{J} = \mathrm{e}^{-2\mathrm{i}\mathrm{Tr}\beta\gamma_5}. \tag{3.28}$$

This determinant diverges so we need to introduce a regulator. Let us first separate the divergent part of the trace

$$\mathrm{Tr}\,\beta\gamma_5 = \mathrm{Tr}'\int \mathrm{d}^4x\langle x|\beta\gamma_5\rangle, \tag{3.29}$$

where the primed trace is over internal degrees of freedom. To regularize the integral we rewrite it as

$$\mathrm{Tr}\,\beta\gamma_5 = \lim_{M\to\infty}\mathrm{Tr}'\int \mathrm{d}^4x\langle x|\beta\gamma_5\mathrm{e}^{-\left(\frac{\not{D}}{M}\right)^2}|x\rangle. \tag{3.30}$$

The square in the regulator can be expanded

$$\left(\frac{\not{D}}{M}\right)^2 = \frac{1}{M^2}\left(\partial_\mu\partial^\mu + \frac{1}{4}g_3\lambda^a\sigma^{\mu\nu}F_{\mu\nu}^a\right) \equiv \frac{1}{M^2}\left(D^2 + g_3\sigma\cdot F\right). \tag{3.31}$$

Here the definition of $\sigma \cdot F$ is given implicitly. To calculate the trace of the regulator we introduce the complete set of states

$$\begin{aligned}\mathrm{Tr}\,\beta\gamma_5 &= \lim_{M\to\infty}\mathrm{Tr}'\int \mathrm{d}^4x\frac{\mathrm{d}^dp}{(2\pi)^{d/2}}\frac{\mathrm{d}^dp'}{(2\pi)^{d/2}}\langle x|p\rangle\langle p|\beta\gamma_5\mathrm{e}^{-\left(\frac{\not{D}}{M}\right)^2}|p'\rangle\langle p'|x\rangle \\ &= \lim_{M\to\infty}\mathrm{Tr}'\int \mathrm{d}^4x\frac{\mathrm{d}^dp}{(2\pi)^d}\beta\gamma_5\mathrm{e}^{-\left((\mathrm{i}p_\mu+D_\mu)^2+g_3\sigma\cdot F\right)/M^2} \\ &= \lim_{M\to\infty}\mathrm{Tr}'\int \mathrm{d}^4x\frac{\mathrm{d}^dp}{(2\pi)^d}\beta\gamma_5\mathrm{e}^{p^2/M^2}\mathrm{e}^{-(D^2+g_3\sigma\cdot F+2ip\cdot D)/M^2}.\end{aligned} \tag{3.32}$$

We will now Taylor expand the right-hand side up to $\mathcal{O}(1/M^4)$ keeping in mind that each factor of p^2 is of order $\mathcal{O}(M^2)$

$$
\begin{aligned}
\mathrm{Tr}\,\beta\gamma_5 = &\lim_{M\to\infty} \mathrm{Tr}' \int \mathrm{d}^4x \frac{\mathrm{d}^d p}{(2\pi)^d} \beta\gamma_5 \mathrm{e}^{\frac{p^2}{M^2}} \Big[1 - \frac{1}{M^2}\left(D^2 + g_3\sigma\cdot F\right) \\
&+ \frac{1}{2M^4}\left(\left\{D^2 + g_3\sigma\cdot F\right\}^2 - 4p\cdot D\right) + \frac{2}{3M^6} \\
&\times \left(p\cdot Dp\cdot D\left\{D^2 + g_3\sigma\cdot F\right\} + p\cdot D\left\{D^2 + g_3\sigma\cdot F\right\}p\cdot D\right. \\
&+ \left.\left\{D^2 + g_3\sigma\cdot F\right\}p\cdot Dp\cdot D\right) + \frac{4}{M^8}(p\cdot D)^4\Big] \\
= &\lim_{M\to\infty} \mathrm{Tr}'\mathrm{i} \int \mathrm{d}^4x \beta\gamma_5 \frac{iM^4}{(4\pi)^2} \times \Big[1 - \frac{g_3\sigma\cdot F}{M^2} \\
&+ \frac{1}{M^4}\left(\frac{1}{2}(g_3\sigma\cdot F)^2 + \frac{1}{12}\left[D_\mu, D_\nu\right]^2 + \frac{1}{6}\left[D_\mu, \left[D_\mu, \sigma\cdot F\right]\right]\right)\Big].
\end{aligned}
\tag{3.33}
$$

The only trace that survives is the one with a product of two σ. Recalling that

$$
\mathrm{Tr}\,\gamma_5\sigma^{\mu\nu}\sigma^{\alpha\beta} = -4\mathrm{i}\epsilon^{\mu\nu\alpha\beta}, \tag{3.34}
$$

it is then easy to take the trace over the internal degrees of freedom to show that

$$
\mathcal{J} = \exp\left[-\mathrm{i}\int \mathrm{d}^4x \beta \frac{3\alpha S}{8\pi} F^{\mu\nu}\tilde{F}_{\mu\nu}\right]. \tag{3.35}
$$

We therefore have

$$
\partial_\mu J_5^\mu = \frac{3\alpha_s}{8\pi} F^{\mu\nu}\tilde{F}_{\mu\nu}. \tag{3.36}
$$

This effective operator generates the amplitude given in equation (3.15) which was derived using operator techniques and dimensional regularization.

3.1.3 Baryon and lepton number violation

At this stage we have rederived the chiral anomaly using functional techniques and taking into account the gauge symmetries of the Standard Model. The SU(2) gauge bosons couple only to left-handed particles so the chiral anomaly interferes with lepton and baryon number conservation. Consider the classically conserved currents

$$
J^\mu_{L,R} = \frac{1}{2}\bar{\psi}\gamma^\mu(1 \mp \gamma^5)\psi. \tag{3.37}
$$

Let us first couple the left-handed current to the Standard Model. Following the procedure as before we consider the term $W[a_\mu + \partial_\mu\beta, A_\mu, B_\mu, G_\mu]$ where $L = L_{\mathrm{SM}} + (a_\mu + \partial_\mu\beta)J^\mu_L$. This can be related to the term $W[a_\mu, A_\mu, B_\mu, G_\mu]$ by the transformations

$$
\psi' = (1 - \mathrm{i}\beta\gamma_5)\psi \sim \mathrm{e}^{-\mathrm{i}\beta\gamma_5}\psi \tag{3.38}
$$

$$\bar{\psi}' = \bar{\psi}(1 - \mathrm{i}\beta\gamma_5) \sim \bar{\psi}\mathrm{e}^{-\mathrm{i}\beta\gamma_5}. \tag{3.39}$$

We then can use the method outlined before to derive an expression for $\partial_\mu J_L^\mu$ with associated regulator $\exp\{[(\partial_\mu + gA_\mu + g'B_\mu + g_s G_\mu)\gamma^\mu]^2/M^2\} \equiv \exp[\not{D}_L^2/M^2]$. Similarly we can couple the right-handed current to our Lagrangian. The transformation required to bring us back to our original W function is

$$\psi' = (1 + \mathrm{i}\beta\gamma_5)\psi \sim \mathrm{e}^{+\mathrm{i}\beta\gamma_5}\psi \tag{3.40}$$

$$\bar{\psi}' = \bar{\psi}(1 + \mathrm{i}\beta\gamma_5) \sim \bar{\psi}\mathrm{e}^{+\mathrm{i}\beta\gamma_5}. \tag{3.41}$$

Our Jacobian will acquire an extra minus sign. This time the associated regulator is $\exp\{[(\partial_\mu + gA_\mu + g_s G_\mu)\gamma^\mu]^2/M^2\} \equiv \exp[\not{D}_R^2/M^2]$. Following the previous calculation one finds for quarks

$$\partial_\mu [J_L^\mu - J_R^\mu] \equiv \partial_\mu J_5^\mu = \frac{3\alpha_S}{8\pi} G^{\mu\nu}\tilde{G}_{\mu\nu} \tag{3.42}$$

$$\partial_\mu J_L^\mu = \frac{3\alpha_W}{8\pi} B^{\mu\nu}\tilde{B}_{\mu\nu}. \tag{3.43}$$

When one performs the calculation for leptons there is only the anomalous left-handed current which is equal to one third the above equation since one does not perform a trace over colour. Noting that the baryon number of a quark is 1/3 we can write the exact symmetry

$$\partial_\mu B^\mu = \partial_\mu L^\mu. \tag{3.44}$$

3.2 The Chern–Simons form, baryon number violation, and the winding number

We have now demonstrated that both lepton and baryon number are violated in the Standard Model through these strange anomalous currents. In this subsection we will try and find the sort of field configuration that violates baryon and lepton number. This is one part of the background theory which can quickly get unnecessarily formal, so we will be taking an approach to make things only as formal as is useful to understanding the subject at the level needed to do research in this field[2]. Let us first make some manipulations to our effective action. Recall that

[2] We do not follow any particular reference as they often lack details in the areas we need and are too formal or too detailed in areas orthogonal to this analysis. Nonetheless, some useful resources for more information are [10–12].

$$
\begin{aligned}
\partial_\mu L^\mu = \partial_\mu B^\mu &= \frac{g^2}{32\pi^2}\int \mathrm{d}^4x\epsilon^{\mu\nu\alpha\beta}\,\mathrm{Tr}\big[F_{\mu\nu}F_{\alpha\beta}\big] \\
&= \frac{g^2}{32\pi^2}\int \mathrm{d}^4x\epsilon^{\mu\nu\alpha\beta}\partial_\mu\,\mathrm{Tr}\left[\partial_\mu A_\nu^a - \partial_\nu A_\mu^a + f^{abc}A_\mu^b A_\nu^c\right] \\
&\times\left[\partial_\alpha A_\beta - \partial_\beta A_\alpha^a + f^{ade}A_\alpha^d A_beta^e\right] \\
&= \frac{g^2}{8\pi^2}\int \mathrm{d}^4x\epsilon^{\mu\nu\alpha\beta}\partial_\mu\,\mathrm{Tr}\left[A_\nu\partial_\alpha A_\beta - \frac{2\mathrm{i}}{3}A_\nu A_\alpha A_\beta\right] \\
&+ f^{abc}f^{ade}\epsilon^{\mu\nu\alpha\beta}\,\mathrm{Tr}\Big[\big[A_\mu^a,\, A\nu^b\big]\big[A_\alpha^d A_\beta^e\big]\Big] \\
&= \frac{g^2}{8\pi^2}\int \mathrm{d}^4x\epsilon^{\mu\nu\alpha\beta}\partial_\mu\,\mathrm{Tr}\left[A_\nu\partial_\alpha A_\beta - \frac{2\mathrm{i}}{3}A_\nu A_\alpha A_\beta\right] \\
&= \frac{g^2}{8\pi^2}\int_{S^3}\mathrm{Tr}\left[A\wedge\mathrm{d}A + \frac{1}{3}A\wedge A\wedge A\right].
\end{aligned}
\tag{3.45}
$$

Note we used the Jacobi identity in the above derivation. Also in the last line we have used the divergence theorem and wrote the result in a compact notation using wedge products. If you are unfamiliar with this notation rest assured we will use them sparingly. The integrand here is known as the 'Chern–Simons action' as it is the surface integral of the Chern–Simons form [13]. The field configurations that will be relevant are finite-energy solutions, so we require them to be well-behaved at infinity

$$
A_\mu \to \mathrm{i}g\partial_\mu g^{-1} + \mathcal{O}\left(\frac{1}{r^2}\right). \tag{3.46}
$$

That is, we require that in some gauge the field vanishes at a rate faster than $1/r^2$ up to 'pure gauge' contributions. Field configurations that do not satisfy the finite-energy condition make no contribution to the functional integral so they can be ignored. We note that the integral of the Chern–Simons form is the volume integral of a total divergence. We can then use Gauss' divergence theorem to turn this into an integral over the surface at spatial infinity. If we enter the pure gauge part into equation (3.45) we obtain

$$
\begin{aligned}
\partial_\mu L^\mu &= \frac{g^2}{8\pi^2}\int \mathrm{d}S_\mu\epsilon^{\mu\nu\alpha\beta}\,\mathrm{Tr}\left[\mathrm{i}^2 g\partial_\nu g^{-1}\big(\partial_\alpha g\partial_\beta g^{-1}\big) - \frac{2\mathrm{i}^4}{3}g\partial_\mu g^{-1}g\partial_\alpha g^{-1}g\partial_\beta g^{-1}\right] \\
&= \frac{g^2}{8\pi^2}\int \mathrm{d}S_\mu\epsilon^{\mu\nu\alpha\beta}\,\mathrm{Tr}\left[-g\partial_\nu g^{-1}\big(\partial_\alpha g(gg^{-1})\partial_\beta g^{-1}\big) - \frac{2}{3}g\partial_\mu g^{-1}g\partial_\alpha g^{-1}g\partial_\beta g^{-1}\right] \\
&= \frac{g^2}{8\pi^2}\int \mathrm{d}S_\mu\epsilon^{\mu\nu\alpha\beta}\,\mathrm{Tr}\left[g\partial_\mu g^{-1}g\partial_\alpha g^{-1}g\partial_\beta g^{-1} - \frac{2}{3}g\partial_\mu g^{-1}g\partial_\alpha g^{-1}g\partial_\beta g^{-1}\right] \\
&= \frac{g^2}{24\pi^2}\int \mathrm{d}S_\mu\epsilon^{\mu\nu\alpha\beta}\,\mathrm{Tr}\left[g\partial_\mu g^{-1}g\partial_\alpha g^{-1}g\partial_\beta g^{-1}\right],
\end{aligned}
\tag{3.47}
$$

where in the second line we used the identity

$$\partial_x 1 = 0 = \partial_x(gg^{-1}) = (\partial_x g)g^{-1} + g(\partial_x g^{-1}). \tag{3.48}$$

The form of this is very interesting. Consider the volume element for the internal group

$$\mathrm{d}\Omega_V = \frac{1}{24\pi^2}\int \mathrm{Tr}\left[(\mathrm{d}gg^{-1}) \wedge (\mathrm{d}gg^{-1}) \wedge (\mathrm{d}gg^{-1})\right]. \tag{3.49}$$

So the particle current divergence is proportional to the volume integral over the internal group pulled back onto Euclidean space–time. In other words it is proportional to the number of times the field winds around the group at the three-dimensional surface at spatial infinity. If the integral of the Chern–Simons form is non-zero we therefore denote it with a 'non-zero' Chern–Simons number due to the connection with the winding number.

3.3 Winding number and non-abelian gauge groups

The kind of field configuration that violates baryon number conservation is one with a non-zero 'Chern–Simons number', which we argued in the previous section is satisfied if the winding number is non-zero. In this section we will make things a little more concrete. The winding number is a concept from topology. As we only need surface level understanding of the concept we will only give a few examples to build intuition toward the subject. First consider the case where the internal space of a gauge field is a simple circle. Now consider the following mapping:

$$g^\nu(\theta) = \mathrm{e}^{(\mathrm{i}\nu\theta)}. \tag{3.50}$$

The winding number is defined as how often this function winds around the circle. It is easy to show that the following equation is the winding number:

$$\frac{\mathrm{i}}{2\pi}\int_0^{2\pi} \mathrm{d}\theta \mathrm{d}g^{-1}/\mathrm{d}\theta. \tag{3.51}$$

What is obvious also from explicit calculation is that the product of two mappings gives a winding number that is the sum of the winding numbers of each mapping. That is, if the winding number of one function, $f(\theta)$ is ν_1 and the winding number of $g(\theta)$ is ν_2, then the winding number of $f(\theta)g(\theta)$ is $\nu_1 + \nu_2$. This just follows from the properties of exponential functions. However, let us prove this in a way that may seem needlessly complicated for this simple example, but will be useful for the more complicated case we are interested in. First we consider a small change in the winding number

$$\delta g = \mathrm{i}(\delta\lambda). \tag{3.52}$$

It is easy to show that $\delta\nu = 0$,

$$\begin{aligned}\nu + \delta\nu &= \frac{\mathrm{i}}{2\pi}\int_0^{2\pi} \mathrm{d}\theta\,(g + \delta g)\frac{\mathrm{d}(g^{-1} - \delta g)}{\mathrm{d}\theta}\mathrm{d}\theta \\ &= \nu + \frac{1}{2\pi}\int_0^{2\pi}\left(\delta g\frac{\mathrm{d}g^{-1}}{\mathrm{d}\theta} - \mathrm{i}\frac{\mathrm{d}\delta\lambda g^{-1}}{\mathrm{d}\theta}\right) \\ &= \nu + \frac{1}{2\pi}\int_0^{2\pi}\mathrm{d}\theta\frac{\mathrm{d}\delta\lambda}{\mathrm{d}\theta} \\ &= \nu.\end{aligned} \tag{3.53}$$

Thus we have shown that the winding number is invariant under continuous transformations. So we see that the winding number of the product of two functions is just the sum of the two winding numbers. Now consider the case where the gauge field has a local SU(2) symmetry. In the previous section we claimed that the winding number for this group is given by

$$\nu = \frac{1}{24\pi^2}\int \mathrm{d}S_\nu \epsilon^{\nu\mu\alpha\beta}\,\mathrm{Tr}\left[g\partial_\mu g^{-1} g\partial_\alpha g^{-1} g\partial_\beta g^{-1}\right], \tag{3.54}$$

which we will prove to be the winding number in a way completely analogous to the case of a circle. It is straightforward by direct substitution to take a configuration with winding number of unity, $g = \sigma \cdot \mathbf{x}/r$ (where of course $\sigma^4 = 1_2$), and show that it indeed gives one when inserted into the above equation. So we can prove it is the winding numbing by demonstrating that winding number of two maps is the sum of their individual winding numbers using the same techniques we used for the circle. Once again we begin with considering a small change in the winding number. Recall the identity that

$$(\partial_x g)g^{-1} = -g(\partial_x g^{-1}), \tag{3.55}$$

which we can use to write

$$\delta\nu = \int \mathrm{d}\sigma_\nu \epsilon^{\nu\mu\alpha\beta}\,\mathrm{Tr}\left[\partial_\mu g^{-1}\partial_\alpha g\partial_\beta\delta\lambda^a(x)\tau^a\right]. \tag{3.56}$$

This vanishes upon integration by parts. It is then trivial to see that our expression for ν is indeed the winding number as one can then continuously deform our expression for g such that it is equal to one for the upper part of the hypersphere, and deform another mapping of winding number one such that it is equal to one for the lower part of the hypersphere. The winding number of the product of these two mappings is then 2. A similar process can produce a mapping g_ν with winding number ν. So what does this mean? If we consider the SU(2) gauge fields in four dimensional Euclidean space and track their value as we walk around spatial infinity, if the field winds around its internal group at least once in this walk then baryon and lepton number is violated. So to understand the process of baryon number violation we will then seek to understand these topologically interesting processes. Specifically we will find in the following sections that we are interested in lumps of field, from the perspective of four-dimensional Euclidean space, called instantons, and unstable temporally localized lumps called sphalerons.

3.4 Solitons and instantons

In the previous sections we demonstrated that field configurations that violate baryon or lepton number must be topologically non-trivial—that is they must have non-zero Chern–Simons number. In this section we will demonstrate that topologically interesting field configurations generically are associated with quasi-particles.

Let us begin with the simplest example. Consider a simple scalar field in 1 + 1 dimensions with degenerate minima. For simplicity and clarity we will set a bunch of parameters in this theory to one to avoid them cluttering our equations as we only wish to discuss some qualitative features of the theory. The 'uncluttered' version of the Lagrangian for 1 + 1-dimensional scalar field theory is

$$\mathcal{L} = \frac{1}{2}\left(\frac{\mathrm{d}\phi}{\mathrm{d}t}\right)^2 - \frac{1}{2}\left(\frac{\mathrm{d}\phi}{\mathrm{d}x}\right)^2 - \frac{1}{4}(\phi^2 - 1)^2. \tag{3.57}$$

Now consider finite-energy solutions to the action that are stationary in time (that is we set the $\mathrm{d}\phi/\mathrm{d}t$ term to zero). Without putting pen to paper we can already make remarks about the types of solutions we will find. The key to this is knowing that if the solution is finite then the field must take the value of one of the minima at plus and minus spatial infinity. This gives us two types of solution, a field that has the same minima at $\pm\infty$ and a field that goes from one minimum at $-\infty$ to another at $+\infty$. The second class of solutions is topologically non-trivial. Consider a specific example of the second type of solution to the classical equations of motion

$$\phi(x) = \pm\tanh\left[(x - x_0)/\sqrt{2}\right]. \tag{3.58}$$

The energy profile of this solution is a highly localized lump of field. This can be seen if we write the energy density of the solution

$$\epsilon(x) = \frac{1}{2}\,\mathrm{sech}^2\left[(x - x_0)/\sqrt{2}\right] \tag{3.59}$$

which we depict in figure 3.2. This is known as a classical lump or a soliton. It is a quasi-particle whose existence derives from a degeneracy in the ground state. Although the solution given above is stationary, we can easily boost to a moving frame to obtain moving soliton solutions. Recalling that under a Lorentz transformation, $\phi(x) \to \phi(\Lambda x)$, we can write the following class of moving solitons

$$\phi(x, t) = \tanh\left[\frac{1}{\sqrt{2}}\left(\frac{(x - x_0) - vt}{\sqrt{1 - u^2}}\right)\right]. \tag{3.60}$$

In the Standard Model we have a very large degeneracy in the ground state since one can have different winding numbers. Field configurations that are topologically interesting in an analogous way to the soliton described above will also look like

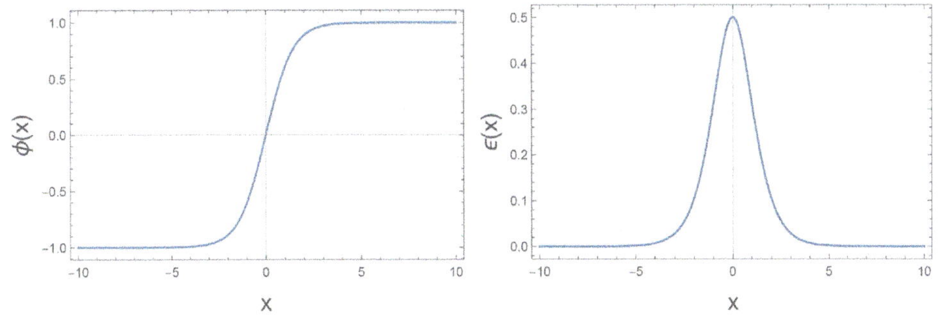

Figure 3.2. Depiction of a soliton in a single spatial dimension for the simple case of two degenerate minima at ±1. The energy profile is highly localized which justifies its status as a quasi-particle.

localized lumps in the field except in all four Euclidean dimensions. These lumps are known as instantons since they are localized in time. If you reverse the Wick rotation, returning to 3 + 1 dimensions, these instantons look like tunneling processes [14]. Having more dimensions and a more complicated degeneracy means we cannot immediately write down the instanton solution as we did with the soliton. In the case where we have SU(2) gauge symmetry, it helps to first show that the instanton is a solution to the classical equations of motion—it is a local minimum of the action. First take the field strength tensor and make use of the trivial identity

$$\begin{aligned}\int \mathrm{d}^4x \ \mathrm{Tr}\, F_{\mu\nu}F^{\mu\nu} &= \int \mathrm{d}^4x \ \mathrm{Tr}\left[\sqrt{F_{\mu\nu}F^{\mu\nu}\tilde{F}_{\alpha\beta}\tilde{F}^{\alpha\beta}}\right] \\ &\geqslant \int \mathrm{d}^4x \ \mathrm{Tr}\left[\left|F_{\mu\nu}\tilde{F}^{\mu\nu}\right|\right].\end{aligned} \tag{3.61}$$

This means that

$$\int \mathrm{d}^4x \ \mathrm{Tr}\, F_{\mu\nu}F^{\mu\nu} \geqslant \frac{8\pi^2}{g^2}|\nu| \tag{3.62}$$

with the equality being satisfied for $F = \tilde{F}$. To construct an ansatz for an instanton we have the conditions that the winding number is 1, the configuration is finite-energy and the solution is a local extrema of the action. For simplicity let us ignore the gauge coupling constant by setting it to 1 and write the ansatz

$$A_\mu = \mathrm{i}f\left(r^2\right)g(x)\partial_\mu g^{-1}(x), \tag{3.63}$$

where f is a radial function that goes to 1 as $r \to \infty$ faster than $1/r^2$. We can substitute this ansatz into the classical equations of motion (or equivalently the equation $F = \tilde{F}$) to find

$$f\left(r^2\right)^2 - f\left(r^2\right) = -r^2 f'\left(r^2\right), \tag{3.64}$$

which is satisfied for

$$f(r^2) = \frac{r^2}{r^2 + \rho^2}. \tag{3.65}$$

The instanton is then [15, 16]

$$\begin{aligned} A_\mu(x) &= g^{-1}(x-a)\left[\partial_\mu g(x-a)\right]\frac{|(x-a)|^2}{|(x-a)|^2 + \rho^2} \\ \phi(x) &= 1 + \frac{\rho^2}{(x-a)^2}. \end{aligned} \tag{3.66}$$

That $A_\mu(x)$ winds around the group once is clear from the fact that it is a radial function times the pure gauge form for the $\nu = 1$ gauge configuration. This can also be seen by explicitly putting the above solution into the Chern–Simons form (3.47). The above instanton tunnels between two neighboring vacua and in the process produces three leptons and three baryons. If all particles are Standard Model this means nine quarks and three leptons. Calculating the energy density of this solution once again shows a clump of field highly localized in space and time, as can been seen in figure 3.3.

The amplitude for an instanton rate at zero temperature can be found by simply inserting the instanton solution into the path integral and making a saddle-point approximation. Doing so one finds that the instanton amplitude is proportional to $\exp[-8\pi^2/g^2] \sim 10^{-173}$ [14] which is an incredible suppression! So at zero temperature it cannot possibly be important to explain the baryon asymmetry. Finally let us see how the ground state of the Standard Model can be thought of as a periodic potential [17–19]. Let an eigenstate of definite winding number be denoted by $|n\rangle$.

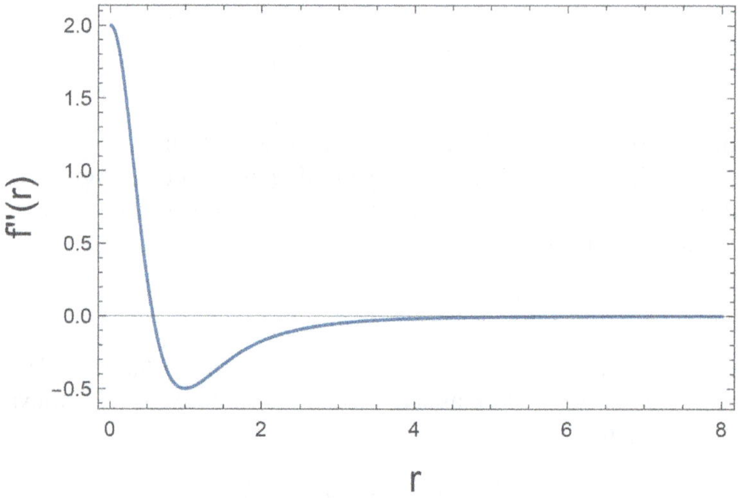

Figure 3.3. The energy profile of an instanton as a function of r^2 is also highly localized. This is the motivation behind thinking of the instanton as a quasi-particle in Euclidean space. Here the parameter $\rho = 1$.

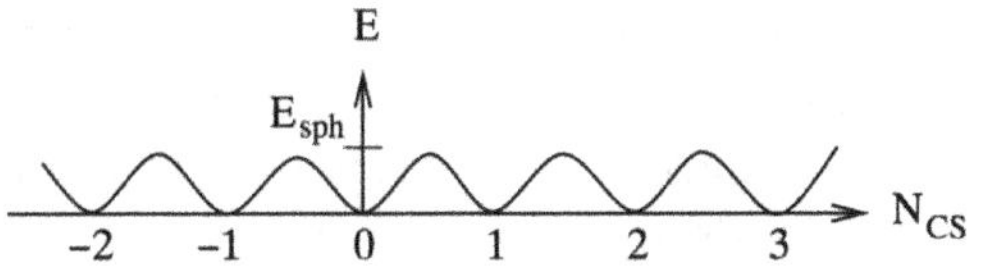

Figure 3.4. Standard Model ground state is actually periodic with a potential height of E_{sph}. Each degenerate minimum corresponds to a different Chern–Simons number. A shift to the right from one minimum to another is the instanton, whereas the local maxima are the sphalerons. Each degenerate ground state has a different Chern–Simons number, N_{CS}. Instantons move one to the right from one ground state to another with a higher Chern–Simons number whereas anti-instantons move one to the left.

The conjugate variable to n will then be a phase variable. An eigenstate of phase can be written as

$$|\theta\rangle = \sum_{n=-\infty}^{\infty} \mathrm{e}^{in\theta}|n\rangle. \tag{3.67}$$

Let us consider the term $\langle\theta'|\exp[-HT]|\theta\rangle$. We of course have calculated the term $\langle n+1|\exp[-HT]|n\rangle - \langle n+1|HT|n\rangle \approx AT\exp[-8\pi^2/g^2]$. We can make the approximation where we ignore multiple tunneling events and consider only single tunneling transitions. Defining $E(\theta)\delta(\theta-\theta') = \langle\theta'|H|\theta\rangle$ we have

$$E(\theta) = E_0 - 2A\exp\left[\frac{-8\pi^2}{g^2}\right]\cos\theta. \tag{3.68}$$

The ground state can be approximated as a periodic potential with a potential height proportional to the tunneling amplitude as depicted in figure 3.4.

Let us conclude the section by summarizing what we have learned so far about baryon number violation in the Standard Model:

- Classical conservation laws can be anomalously broken.
- Triangle anomalies in the Standard Model result in an anomalous violation of both baryon and lepton number conservation.
- The field configurations that anomalously violate baryon/lepton number are ones with a non-zero Chern–Simons number. This means they wrap around the internal space at the surface at spatial infinity in Euclidean space.
- Fields with integer winding number at infinity are solutions to the classical equations of motion.
- Fields with integer winding number look like localized lumps in Euclidean space.

3.5 The sphaleron

Let us now try and bootstrap another topologically interesting quasi-particle that is a good candidate for baryon number violation. Specifically, the quasi-particle we will discuss is the sphaleron whose etymology is from the Greek word that means to 'go down', because it is an unstable particle [20, 21]. In the previous section we presented the instanton with a winding and Chern–Simons number of 1. It turns out that the instanton is less important than the sphaleron for our purposes because sphaleron rates become unsuppressed at high temperatures. As such we give it far closer attention. Physically, the instanton corresponds to tunneling through a barrier whereas a sphaleron corresponds to classically passing over the barrier. A sphaleron is an unstable particle corresponding to an instantaneous moment in time and is a solution to the classical equations of motion. It is due to its importance in the logical framework of this field that we give it such deep attention. It should be noted that we only focus on the lowest energy electroweak sphaleron before touching on the SU(3) sphaleron, since the lowest energy sphaleron will dominate. Of course we note that there is no formal proof that we are aware of that the sphaleron we present is indeed the lowest energy but it is widely believed to be so. To begin our bootstrap let us start with insisting that the Higgs field, which is an $SU(2)_L$ doublet, is well-behaved at infinity. Well-behaved in this case means the covariant derivative must vanish

$$D_\mu \phi \to 0 \text{ as } r \to \infty. \tag{3.69}$$

This gives the condition

$$\partial_\mu \phi \Big|_\infty = igA_\mu^a \tau^a \Big|_\infty \phi \Big|_\infty. \tag{3.70}$$

In this case we keep the gauge coupling constant as we will eventually calculate the sphaleron rate. We want $\partial\phi/\partial t = 0$ and we want a topologically non-trivial solution. So let us set $A_0|_\infty$ to zero along with $\mathrm{d}\phi/\mathrm{d}t|_\infty$ and map the space–time surface onto the internal space by setting A_i^∞ to be proportional to the right invariant one form—also called the Maurer–Cartan form

$$\phi^\infty = \Omega \begin{pmatrix} 0 \\ v/\sqrt{2} \end{pmatrix} \tag{3.71}$$

and

$$A_i^\infty = \frac{-\mathrm{i}}{g}(\partial_i \Omega)\Omega^{-1}. \tag{3.72}$$

We will use Ω instead of g here and throughout for an element of the gauge group to avoid confusion with the coupling constant. The Hermitian conjugate of the above is a solution. Both of these solutions can be verified by direct substitution.

This is the first requirement of a sphaleron: that it reduce to the Maurer–Cartan form at infinity in order to be a finite-energy solution. Fields with integer winding number are known to be (multi-)instantons. The only other topologically interesting cases that can potentially violate baryon number are the cases where we have a non-integer and non-zero Chern–Simons number. It will turn out that sphalerons have a Chern–Simons number of 1/2. However, the Chern–Simons number was shown in the previous section to be related to the winding number, which is forced to be an integer. So it is something of a surprise that a non-integer Chern–Simons number is possible given its close relationship to the winding number. Indeed the

Chern–Simons number of any *vacuum* state must be an integer so a non-integer Chern–Simons number means it must continuously deform from one vacuum state to another.

To complete our bootstrap we explicitly make use of the Pauli matrices and make our fields cover the internal group as we cover the spatial sphere at $r = \infty$ while passing through zero somewhere in space via use of a radial function. Therefore a suitable ansatz is

$$\phi^{\infty} = h(r)\Omega\begin{pmatrix} 0 \\ v/\sqrt{2} \end{pmatrix} \tag{3.73}$$

and

$$A_i^{\infty} = \frac{-\mathrm{i}f(r)}{g}(\partial_i\Omega)\Omega^{-1} \tag{3.74}$$

with

$$\Omega = \vec{\sigma} \cdot \frac{\vec{x}}{r}. \tag{3.75}$$

Also $f \equiv f(r)$ and $h(r)$ are radial functions chosen to solve the equations of motion subject to the boundary conditions $h(0) = f(0) = 0$ and $h(\infty) = f(\infty) = 1$. Let us now complete the motivation for studying closely this quasi-particle over the instanton, by demonstrating that these processes become unsuppressed when electroweak symmetry is restored at high temperature. We apologize to the reader for slightly skipping ahead in the flow of our logical argument by using a small amount of finite temperature field theory from the later parts of this primer. The reader should regard the form of the next few equations as divinely given and then return to this section after reading the later sections on finite temperature quantum field theory (QFT). The sphaleron rate can be found by substituting the above solution into the finite temperature version of the path integral. The high temperature sphaleron rate is given by

$$\Gamma_{\rm sph} = \frac{\omega_-}{2\pi}(\mathcal{N}V)_{\rm rot}(\mathcal{N}V)_{\rm tr}\kappa \mathrm{e}^{\Delta E_{\rm sph}/T}, \tag{3.76}$$

where $\Delta E_{\rm sph}/T = S_3(\phi_{\rm sph})$, where

$$S_3\left[\phi_{\rm sph}\right] = \frac{1}{T}\int \mathrm{d}^3x\, |D_i\phi|^2 + \frac{1}{4}W_{ij}^a W_{ij}^a + V(\phi). \tag{3.77}$$

Here W_{ij} are the field strength tensors for A_i, $W_{ij} \equiv \partial_i A_j - \partial_j A_i$. With some effort we can reduce this to

$$S_3\left[\phi_{\rm sph}\right] \equiv \frac{\Delta E_{\rm sph}}{T} = \frac{4\pi v(T)}{gT}B\left(\frac{\lambda}{g^2}\right) \tag{3.78}$$

with

$$B\left(\frac{\lambda}{g^2}\right) = \int_0^\infty \mathrm{d}r \left[4\left(\frac{\mathrm{d}f}{\mathrm{d}r}\right)^2 + \frac{8}{r^2}f^2(1-f)^2 \right. \\ \left. + \frac{r^2}{2}\left(\frac{\mathrm{d}h}{\mathrm{d}r}\right)^2 h^2(1-f)^2 + \frac{1}{4}\frac{\lambda}{g^2}r^2\left(1-h^2\right)^2\right], \tag{3.79}$$

where $r \equiv gv(T)|x|^2$. Note that when $v(T) = 0$ the rate is no longer exponentially suppressed. We will show in later chapters that electroweak symmetry is indeed restored at very high temperatures, meaning that at some stage in our cosmic history sphalerons, and therefore baryon number violation, were unsuppressed! This is the key feature of sphalerons which makes them attractive as a source of baryon number violation at very high temperature when electroweak symmetry is restored. There is one remaining thing we have to show to conclude our discussion of sphalerons in terms of showing that we indeed have a non-zero Chern–Simons number. Recall that the Chern–Simons form is

$$\mathrm{Tr}\left[W \wedge \mathrm{d}W + \frac{2}{3}W \wedge W \wedge W\right] \tag{3.80}$$

where for $A_0 = 0$, the above can be written

$$\mathrm{Tr}\,[W \wedge \mathrm{d}W] = \epsilon_{ijk}W_i\left(\partial_j W_k - \partial_j W_j\right) = -\epsilon_{ijk}\left(\partial_i W_j - \partial_j W_i\right)W_k \\ \mathrm{Tr}\,[W \wedge W \wedge W] = -\,\epsilon_{ijk}\epsilon_{abc}W_i^a W_j^b W_k^c \tag{3.81}$$

and the Chern–Simons number for the sphaleron is

$$Q(\text{Sphaleron}) = \frac{g^2}{32\pi^2}\int_{t=t_0} \mathrm{d}^3x K^0, \tag{3.82}$$

where K^0 is the Chern–Simons form. Recall our field configuration,

$$W_\mathrm{i}^a\sigma^a = -\frac{2\mathrm{i}f}{g}\Omega^{-1}\mathrm{d}\Omega \tag{3.83}$$

where Ω is given above and $\mathrm{d}\Omega$ is the external derivative $\frac{\partial\Omega}{\partial x_i}\mathrm{d}x_i$. Remarkably the form of f will not affect the Chern–Simons number as long as the boundary conditions given above are met. It is easy to see that the above equation reduces to

$$W_\mathrm{i}^a\mathrm{d}x_i = -\frac{2f}{g}\epsilon_{iab}x_b. \tag{3.84}$$

We cannot calculate the Chern–Simons number in this form as the derivation of the Chern–Simons number assumes that our field goes to zero faster than $1/r$. Let us therefore perform a gauge transformation $U = \exp[\frac{\mathrm{i}}{2}\Theta(r)\sigma \cdot \frac{x}{r}]$, where $\Theta(r)$ is a radial function that goes to zero at the spatial origin. Recall that a gauge transformation for non-abelian fields has the form

$$gW_{\rm i}^{a}\sigma_a \to U^{\dagger}\left(\frac{-2f}{r^2}\epsilon_{aik}x_k\sigma_a\right)U + {\rm i}\left(\partial_{\rm i}U^{-1}\right)U. \tag{3.85}$$

It is straightforward to show that in this gauge the sphaleron solution looks like

$$gW_i^a = A\epsilon_{iab}x_b + B\left(\delta_{ia}r^2 - x_ix_a\right) + \Theta'\frac{x_ix_a}{r^2} \equiv \mathcal{A}_{ai} + \mathcal{B}_{ai} + \mathcal{C}_{ai}, \tag{3.86}$$

where the prime denotes a radial derivative and

$$\begin{aligned} A \equiv A(r) &= \frac{1}{r^2}\left[\left(1 - 2f(r)\right)\cos(\Theta) - 1\right] \text{ and} \\ B \equiv B(r) &= \frac{1}{r^3}\left[\left(1 - 2f(r)\right)\sin(\Theta)\right]. \end{aligned} \tag{3.87}$$

Such abbreviations may be disorientating but we will continue to use them throughout because it becomes far too cumbersome to write out every term explicitly. The sphaleron now has the correct asymptotic behavior. Since $W^a{}_i$ is spherically symmetric the integral over angular variables in the calculation is trivial allowing us to rewrite the Chern–Simons number of the sphaleron as

$$Q(\text{Sphaleron}) = \frac{g^2}{8\pi}\int_{t=t_0} r^2 {\rm d}rK^0. \tag{3.88}$$

Using above equations and verifying this with equation (3.45) we can write

$$K^0 = -\left(\epsilon_{ijk}F_{ij}^aW_k^a + \frac{2g}{3}\epsilon_{ijk}\epsilon_{abc}W_i^aW_j^bW_k^c\right). \tag{3.89}$$

Let us first calculate the first term in the Chern–Simons density. Due to our gauge transformation it consists of three terms

$$\epsilon_{ijk}F_{ij}^aW_k^a \equiv (A_{ak} + B_{ak} + C_{ak})W_k^a, \tag{3.90}$$

which in turn are defined by the three terms we defined before $\mathcal{A}_{ia}$, $\mathcal{B}_{ia}$ and $\mathcal{C}_{ia}$ as follows

$$A_{ak} \equiv \epsilon_{ijk}\left[\partial_i\mathcal{A}_{ja} - \partial_j\mathcal{A}_{ia}\right] \text{ etc.} \tag{3.91}$$

With a bit of effort one can simplify the expressions for (A_{ak}, B_{ak}, C_{ak}), reducing them down to the simple form

$$\begin{aligned} gA_{ak} &= \frac{2}{r}[r^2\delta_{ka} - x_kx_a]A' + 4A\delta_{ka} \\ gB_{ak} &= 2\epsilon_{iak}x_i[B'r + 3B] \end{aligned} \tag{3.92}$$

$$gC_{ak} = -2\epsilon_{iak}x^i\Theta'. \tag{3.93}$$

We now contract the above with $W^a{}_k$ which gives a sum of nine terms It is easy to show that these terms are

$$\begin{array}{lll} g^2A_{ka}\mathcal{A}_{ka} = 0 & g^2A_{ka}\mathcal{B}_{ka} = 4BA'r^3 + 8ABr^2 & g^2A_{ka}\mathcal{C}_{ka} = 4A\Theta' \\ g^2B_{ka}\mathcal{A}_{ka} = -4[B'r + 3B]Ar^2 & g^2B_{ka}\mathcal{B}_{ka} = 0 & g^2B_{ka}\mathcal{C}_{ka} = 0 \\ g^2C_{ka}\mathcal{A}_{ka} = 4A\Theta'r^2 & g^2C_{ka}\mathcal{B}_{ka} = 0 & g^2C_{ka}\mathcal{C}_{ka} = 0. \end{array} \tag{3.94}$$

So we have, in summary, the first term in the Chern–Simons density calculated as

$$\frac{g^2r^2}{8\pi}\epsilon_{ijk}F^a_{ij}W^a_k = \frac{1}{2\pi}BA'r^5 + \frac{1}{\pi}ABr^4 + \frac{1}{2\pi}r^4A[B'r + 3B] + \frac{1}{\pi}A\Theta'r^4. \tag{3.95}$$

Next we have the term $g^2\epsilon_{ijk}\epsilon_{abc}W_i^aW_j^bW_k^c$. Let us first calculate the product of two sphalerons contracted with the permutation symbols

$$\epsilon_{ijk}\epsilon_{abc}W_i^aW_j^b = \epsilon_{abc}\epsilon_{ijk}[\mathcal{A}_{ai} + \mathcal{B}_{ai} + \mathcal{C}_{ai}]\left[\mathcal{A}_{bj} + \mathcal{B}_{bj} + \mathcal{C}_{bj}\right]. \tag{3.96}$$

It is useful to organize the terms into a matrix

$$\begin{aligned} g^2\epsilon_{ijk}\epsilon_{abc}&\begin{pmatrix} \mathcal{A}_{ai}\mathcal{A}_{bj} & \mathcal{A}_{ai}\mathcal{B}_{bj} & \mathcal{A}_{ai}\mathcal{C}_{bj} \\ \mathcal{B}_{ai}\mathcal{A}_{bj} & \mathcal{B}_{ai}\mathcal{B}_{bj} & \mathcal{B}_{ai}\mathcal{C}_{bj} \\ \mathcal{C}_{ai}\mathcal{A}_{bj} & \mathcal{C}_{ai}\mathcal{B}_{bj} & \mathcal{C}_{ai}\mathcal{C}_{bj} \end{pmatrix} \\ &= \begin{pmatrix} 2x_kx_cA^2 & 0 & -\Theta'A\epsilon_{kbc}x_b \\ 0 & 2x_kx_cB^2r^2 & B\Theta'\left(\delta_{kc}r^2 - x_kx_c\right) \\ -\Theta'A\epsilon_{kbc}x_b & B\Theta'\left(\delta_{kc}r^2 - x_kx_c\right) & 0 \end{pmatrix} \\ &\equiv M. \end{aligned} \tag{3.97}$$

Contracting the last sphaleron gives the following

$$\begin{aligned} g^2\epsilon_{ijk}\epsilon_{abc}W_i^aW_j^bW_k^c &= [(M_{11})_{kc} + (M_{22})_{kc} + 2(M_{23})_{kc} + 2(M_{13})_{kc}] \\ &\quad\times [\mathcal{A}_{kc} + \mathcal{B}_{kc} + \mathcal{C}_{kc}] \\ &= [\mu_{kc} + \nu_{kc} + \tau_{kc}][\mathcal{A}_{kc} + \mathcal{B}_{kc} + \mathcal{C}_{kc}], \end{aligned} \tag{3.98}$$

where we have again used the abbreviation

$$\mu_{kc} \equiv \{(M_{11})_{kc} + (M_{22})_{kc}\}, \quad \nu_{kc} \equiv 2(M_{23})_{kc}, \quad \tau_{kc} \equiv 2(M_{13})_{kc}. \tag{3.99}$$

We then have

$$\begin{pmatrix} \mu_{kc}\mathcal{A}_{kc} & \mu_{kc}\mathcal{B}_{kc} & \mu_{kc}\mathcal{C}_{kc} \\ \nu_{kc}\mathcal{A}_{kc} & \nu_{kc}\mathcal{C}_{kc} & \nu_{kc}\mathcal{C}_{kc} \\ \tau_{kc}\mathcal{A}_{kc} & \tau_{kc}\mathcal{B}_{kc} & \tau_{kc}\mathcal{C}_{kc} \end{pmatrix} = \begin{pmatrix} 0 & 0 & 2\Theta' r^2\left[A^2 + r^2B^2\right] \\ 0 & 4B^2r^4\Theta' & 0 \\ 4\Theta' A^2 r^2 & 0 & 0 \end{pmatrix}, \tag{3.100}$$

so we can therefore write

$$\frac{g^2r^2}{24\pi}\epsilon_{ijk}\epsilon_{abc}W_i^aW_j^bW_k^c = \frac{1}{4\pi}\Theta' r^4A^2 + \frac{1}{4\pi}r^6B^2\Theta'. \tag{3.101}$$

In total we have

$$\begin{aligned} Q = -\frac{1}{4\pi}\int_{t=t_0} \mathrm{d}r\big[&2BA'r^5 + 4ABr^4 - 2r^4A[B'r + 3B] \\ &+ 4A\Theta' r^2 + 2\Theta' r^4A^2 + 2r^6B^2\Theta'\big]. \end{aligned} \tag{3.102}$$

Simplifying the above equation we can then can the following solvable integral for the Chern–Simons number

$$\begin{aligned} Q(\text{sphaleron}) &= \frac{1}{2\pi}\int_{\Theta(0)}^{\Theta(\infty)} \mathrm{d}\Theta[\Theta - \cos\Theta] \\ &= \frac{1}{2} \end{aligned} \tag{3.103}$$

as required.

So let us summarize. The sphaleron is an unstable quasi-particle due to localization in time and it must continuously go between vacuum states. It is a finite-energy solution so it is well-behaved at the spatial boundary. The simplest solution is then derived by mapping the external space onto the internal space and coupling to a radial function that gives the desired properties and is determined by insisting that the sphaleron be a solution to the classical equations of motion. Such a sphaleron would be a local extremum of the action for any gauge group with an SU(2) subgroup. It has the nice property that the sphaleron rate is unsuppressed when electroweak symmetry is restored. It also has the unusual property that the Chern–Simons number is fractional and shown to be to equal 1/2. That it is non-zero means it facilitates baryon number non-conservation.

References

[1] Coon S A and Holstein B R 2002 Anomalies in quantum mechanics: the $1/r^2$ potential *Am. J. Phys.* **70** 513

[2] Holstein B R 2014 Understanding an anomaly *Am. J. Phys.* **82** 591

[3] Adler S 1969 Axial-vector vertex in spinor electrodynamics *Phys. Rev.* **177** 2426

[4] Bell J S and Jackiw R 1969 A PCAC puzzle: $\pi 0 \rightarrow \gamma\gamma$ in the omega-model *Nuovo Cimento* A **60** 47

[5] Jegerlehner F 2000 Facts of life with gamma (5) arXiv: 0005255(hep-th)

[6] El-Menoufi B K and White G A 2015 The axial anomaly, dimensional regularization and Lorentz-violating QED arXiv: 1505.01754(hep-th)

[7] Ferrari R 2014 Managing γ_5 in dimensional regularization and ABJ anomaly arXiv: 1403.4212

[8] Fujikawa K 1979 Path-integral measure for gauge-invariant fermion theories **42** 18

[9] Donoghue J F, Golowich E and Holstein B R 2014 *Dynamics of the Standard Model* vol 35 (Cambridge: Cambridge University Press)

[10] Rajaraman R 1982 *Solitons and Instantons: An Introduction to Solitons and Instantons in Quantum Field Theory* (North Holland: Amsterdam)

[11] Coleman S 1988 *Aspects of Symmetry: selected Erice lectures* (Cambridge: Cambridge University Press)

[12] Manton N and Sutcliffe P 2004 *Topological Solitons* (Cambridge: Cambridge University Press)

[13] Chern S and Simons J 1974 Characteristic forms and geometric invariants *Ann. Math.* **99** 48–69

[14] 't Hooft G 1976 Symmetry breaking through Bell-Jackiw anomalies *Phys. Rev. Lett.* **37** 8

[15] Wilczek F 1976 Inequivalent embeddings of SU(2) and instanton interactions *Phys. Lett.* B **65** 160

[16] Belavin A A, Polyakov A M, Schwartz A S and Tyupkin Y S 1975 Pseudoparticle solutions of the Yang–Mills equations *Phys. Lett.* B **59** 79

[17] Jackiw R and Rebbi C 1976 Vacuum periodicity in a Yang–Mills quantum theory *Phys. Rev. Lett.* **37** 172

[18] Callan C G, Dashen R F and Gross D J 1976 The structure of the gauge theory vacuum *Phys. Lett.* B **63** 334

[19] Polyakov A M 1997 *Quark Confinement and Topology of Gauge Theories* B **120** 429

[20] Manton N S 1983 Topology in the Weinberg-Salam theory *Phys. Rev.* D **28** 8

[21] Klinkhamer F R and Manton N S 1984 A saddle-point solution in the Weinberg-Salam theory *Phys. Rev.* D **30** 10

Chapter 4

Phase transitions

When the early Universe was at a very high temperature the ground state of the Higgs potential changed [1]. For temperatures well above the weak scale the ground state is $\langle h \rangle = 0$ and electroweak symmetry is restored. As the Universe cools, a phase transition occurs when it becomes energetically favorable for electroweak symmetry to be spontaneously broken. Since sphalerons, a baryon number violating process, are unsuppressed in the high temperature phase, this phase transition becomes of central focus when studying electroweak baryogenesis [2–4].

This phase transition fulfils Sakharov's condition that there must be a departure from equilibrium. However, it turns out that it is insufficient to have an electroweak phase transition. The way the phase transition occurs can qualitatively affect the amount of baryon asymmetry. If the phase transition is second-order or cross-over, the sphalerons which you rely on for baryon number violation work against you by washing out any asymmetry. What is needed to produce a large enough baryon asymmetry is a strongly first-order phase transition which is characterized by a medium of symmetric vacuum with bubbles where electroweak symmetry is broken. Such bubble nucleation occurs when there is a barrier between the new and old minima in the effective potential so that the transition needs to occur by tunneling as depicted in figure 4.1.

If there is a strong source of CP violation then interactions with the bubble wall produce the baryon asymmetry, which is then frozen in as the bubble grows to become the observable Universe. If the suppression of sphalerons within the bubble is strong enough the baryon asymmetry formed in front of its advancing wall is swept up into the new phase and frozen in. The order of the electroweak phase transition within the Standard Model is dependent on the mass of the Higgs where the order of the phase transition becomes second-order and eventually cross-over for a heavier Higgs. For a Higgs mass of 125 GeV the electroweak phase transition is cross-over. The order of the phase transition can be made first-order by additional scalar fields. However, a first-order phase transition is necessary but not sufficient

doi:10.1088/978-1-6817-4457-5ch4

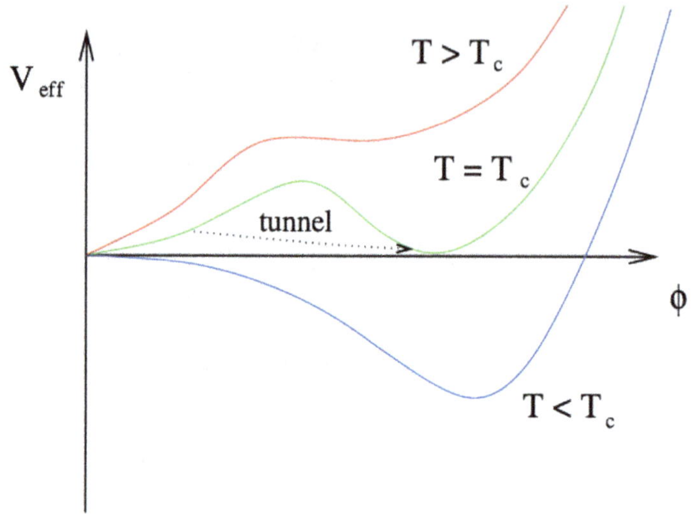

Figure 4.1. Schematic of strongly first-order phase transition. Such a phase transition is characterized by a barrier separating two phases and the barrier persisting when the new phase is lower than the old. In such a case the transition proceeds via tunneling processes. There are three temperatures of interest, the critical temperature where the minima are degenerate (green line), the temperature some time above the critical temperature where the minimum first appears (red line), and the temperature somewhere below the critical temperature where tunneling rates are fast enough that there is at least one critical bubble per Hubble volume (the nucleation temperature).

for sphaleron suppression inside the bubble. We need the sphaleron rate to be very small within the bubble and since the sphaleron rate is suppressed by a factor of $\exp[-v_H(T)/T \times \cdots]$ we can parametrize the strength of the phase transition by $v_H(T_C)/T_C$, where T_C is the critical temperature defined as the temperature at which the vacuum has a degenerate minimum between the origin and the electroweak symmetry breaking phase. Numerically one can derive an approximate condition for baryon particle number conservation of [5]

$$v_H(T_C)/T_C \gtrsim 1. \tag{4.1}$$

Sometime below the critical temperature is the temperature at which a bubble forms that is of a critical size which grows to become the observable Universe (colliding with other critical bubbles in the process). This temperature, known as the nucleation temperature, will be defined as T_N. The crucial quantities we will need to calculate are therefore:

- the critical temperature;
- the order and strength of the phase transition;
- the nucleation temperature;
- the space time bubble wall profile;
- the nucleation rate; and
- the sphaleron rate.

The best estimates of each of these is performed using non-perturbative Monte Carlo methods. However, such methods are so difficult to apply that only a few specific models of electroweak symmetry breaking have been studied using such methods. We therefore focus on finite temperature perturbative techniques. Such techniques provide a reasonable approximation of the more rigorous non-perturbative methods giving qualitative information about the relationship of the above list of quantities on model parameters. However, any calculation involving perturbative finite temperature methods should be assumed to contain a large amount of theoretical uncertainty. The temperature at which the phase transition occurs within the Standard Model, for example, is significantly lower when calculated non-perturbatively compared to the value obtained through perturbative techniques.

The pre-requisite knowledge for performing these calculations is finite temperature QFT and we can progress no further without a review. Finite temperature QFT is a very broad subject so we will limit our discussion to only the key concepts necessary for calculations in this field while keeping things as self-contained as possible. We will assume knowledge of the effective potential and its derivation at zero temperature. After giving a small pedagogical review of finite temperature field theory we will derive the effective potential at finite temperature and calculate the order of the phase transition and its order in the Standard Model. We will then demonstrate the need for the baryon particle number conservation condition to be met before explaining methods of making it strongly first-order through the introduction of new scalar fields. Some issues of gauge invariance will be discussed. Then we will give an overview on how the vacuum decays from a false vacuum to true vacuum, how to calculate the tunneling rate of this process, and the nucleation temperature.

4.1 Closed time path formalism

To describe the particle dynamics during the electroweak phase transition we need to reformulate QFT slightly to handle the out of equilibrium nature of this process. At equilibrium we define states at the infinite past and use the usual machinery of QFT to then predict the 'out states' at the infinite future. Out of equilibrium we do not know the infinite future. What we can do is analyze a system which has been in equilibrium until a point $t = 0$ where the system departs from equilibrium. The in-states can then be evolved to the point where the system departs from equilibrium and we can evolve the system from there [6]. Recall that at finite temperate the density matrix of a system is described as

$$\rho(0) = \frac{e^{-\beta\mathcal{H}}}{\mathrm{Tr}e^{-\beta\mathcal{H}}}, \tag{4.2}$$

where the argument of the density matrix is given to indicate it is at some time $t = 0$. When our system departs from equilibrium, our density matrix acquires a time dependence in the usual way

$$\rho(t) = U(t, 0)\rho(0)U^{\dagger}(t, 0) \tag{4.3}$$

with the time evolution unitary operator defined

$$\begin{aligned} \mathrm{i}\frac{\partial U}{\partial t} &= H(t)U(t, t') \\ U(t, t') &= T\left(\mathrm{e}^{-\mathrm{i}\int_{t'}^{t} \mathrm{d}t'' H(t'')}\right). \end{aligned} \tag{4.4}$$

Note that from the above we obtain the relation that $U(t_1, t_2)^\dagger = U(t_2, t_1)$ and $U(t_1, t_2)U(t_2, t_3) = U(t_1, t_3)$ when $t_1 > t_2 > t_3$. Let the system be in equilibrium up to $t = 0$ meaning that there is no time dependence in the Hamiltonian up to a time where the system goes out of equilibrium (say at the electroweak phase transition) at the time $t = 0$. Note that

$$T\left(\mathrm{e}^{-\mathrm{i}\int_{T-\mathrm{i}\beta}^{T} \mathrm{d}t \mathcal{H}(t)}\right) = T\left(\mathrm{e}^{-\mathrm{i}\int_{T-\mathrm{i}\beta}^{T} \mathrm{d}t \mathcal{H}_0}\right) = \mathrm{e}^{-\beta\mathcal{H}}. \tag{4.5}$$

We can therefore write the equilibrium and non-equilibrium density matrices as

$$\begin{aligned} \rho(0) &= \frac{U(T - \mathrm{i}\beta)}{\mathrm{Tr}\, U(T - \mathrm{i}\beta, T)} \\ \rho(t) &= \frac{U(t, 0)U(T - \mathrm{i}\beta)U(0, t)}{\mathrm{Tr}\, U(T - \mathrm{i}\beta, T)}. \end{aligned} \tag{4.6}$$

The time-dependent expectation value of an operator is defined in the usual way $\langle A(t)\rangle = \mathrm{Tr}\,\rho(t)A$. Let $T < 0$ and $T' \gg 0$. Unitary evolution operators with negative values in both time arguments commute with each other since $\mathcal{H}_0$ commutes with itself. We can write the time-dependent expectation value of an operator as

$$\begin{aligned} \langle A(t)\rangle &= \frac{U(t, 0)U(T - \mathrm{i}\beta)U(0, t)A_0}{\mathrm{Tr}\, U(T - \mathrm{i}\beta, T)} \\ &= \frac{U(t, 0)U(0, T)U(T, 0)U(T - \mathrm{i}\beta)U(0, t)A_0}{\mathrm{Tr}\, U(T - \mathrm{i}\beta, T)} \\ &= \frac{U(t, T)U(T - \mathrm{i}\beta)U(T, t)A_0}{\mathrm{Tr}\, U(T - \mathrm{i}\beta, T)} \\ &= \frac{U(t, T)U(T - \mathrm{i}\beta)U(T, T')U(T', T)U(T, t)A_0}{\mathrm{Tr}\, U(T - \mathrm{i}\beta, T)U(T, T')U(T', T)} \\ &= \frac{U(T - \mathrm{i}\beta, T)U(T, T')U(T', t)AU(t, T)}{\mathrm{Tr}\, U(T - \mathrm{i}\beta, T)U(T, T')U(T', T)}. \end{aligned} \tag{4.7}$$

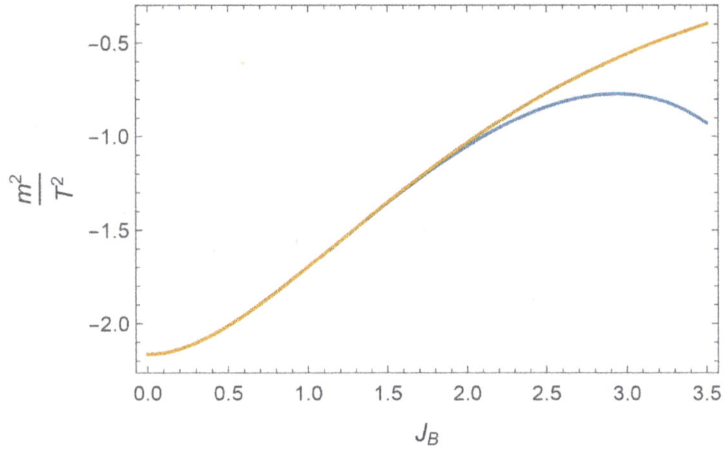

Figure 4.2. $J_B(m^2/T^2)$ compared to its high temperature expansion. The approximation works for $m^2/T^2 \lesssim 2$. The golden line denotes the full thermal function, whereas the blue line denotes the high temperature expansion.

Let us take the limit as $T \to \infty$. Relating the above equation to the type of calculations we wish to perform in QFT (e.g. expectation values of correlation functions), we are performing a different time contour to the standard equilibrium case. Our time contour runs from the infinite past to some time at the present T' where the system is out of equilibrium back into the infinite past before following a path along imaginary time. This is shown in figure 4.2 and is known as the closed time path contour [7–13]. Recall that a generating functional in QFT is given by a source-dependent partition function, say

$$\begin{aligned} Z(\beta) &= \mathrm{Tr} e^{-\beta \mathcal{H}} \\ Z(\beta, \mathcal{J}) &= N \int \mathcal{D}\phi e^{-S_E - \mathcal{J}\phi}. \end{aligned} \tag{4.8}$$

We can then identify the denominator in the last line of equation (4.7) with the generating functional in our out of equilibrium description by suitably adjusting the unitary to include a source term defined along the contour C_i given in figure 4.3

$$\begin{aligned} Z[\beta, \mathcal{J}_c] &= \mathrm{Tr}\, U_{\mathcal{J}_c}(T - i\beta, T) U_{\mathcal{J}_c}(T, T') U_{\mathcal{J}_c}(T', T) \\ &= \int \mathcal{D}\phi e^{i \int_c dt \int d^3x (\mathcal{L} + \mathcal{J}_c \phi)}. \end{aligned} \tag{4.9}$$

Examining the closed time contour, if we wish to form bilinear correlation functions we have four possibilities depending on whether both field operators are on the positive imaginary side of the contour, both are on the negative side, or the two cases where you have one on each. For a scalar field these four cases are

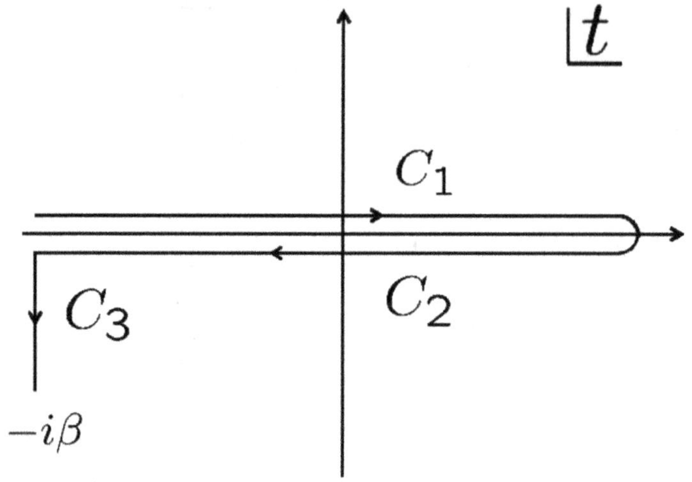

Figure 4.3. The closed time path contour. The contour runs parallel and infinitesimally above the real line from minus infinity to the present, runs parallel and infinitesimally below the real line back to minus infinity, before running parallel to the imaginary axis to $-\mathrm{i}\beta$.

$$\begin{aligned}
\mathrm{i}\Delta^{>}(x, y) &= \langle \phi(x)\phi^{\dagger}(y)\rangle \\
\mathrm{i}\Delta^{<}(x, y) &= \langle \phi^{\dagger}(x)\phi(x)\rangle \\
\mathrm{i}\Delta^{T}(x, y) &= \langle T[\phi(x)\phi^{\dagger}(y)] \\
&= \Theta(x_0 - y_0)\mathrm{i}\Delta^{>}(x, y) + \Theta(y_0 - x_0)\mathrm{i}\Delta^{<}(x, y) \\
\mathrm{i}\Delta^{\bar{T}}(x, y) &= \langle \bar{T}[\phi(x)\phi^{\dagger}(y)] \\
&= \Theta(x_0 - y_0)\mathrm{i}\Delta^{<}(x, y) + \Theta(y_0 - x_0)\mathrm{i}\Delta^{>}(x, y).
\end{aligned} \tag{4.10}$$

To derive the explicit form of these propagators, we first note that $\mathcal{H} \sim a^{\dagger}a$ and that therefore $\mathrm{Tr}[\mathrm{e}^{\beta\mathcal{H}}] = (1 - \mathrm{e}^{\beta\omega})^{-1} = n_B$. This means that $a^{\dagger}a = n_B(\omega_p)\delta^3(p - k)$. Also it will be more convenient to work in momentum space. In the case of spatial isotropy and homogeneity, the momentum space representation version of the propagators is a function of a single momentum variable. The most general form for our propagators can be written as the sum of the zero temperature propagators plus some as yet unspecified momentum conserving, temperature-dependent corrections

$$\begin{pmatrix} \Delta^{T}(p) & \Delta^{<}(p) \\ \Delta^{<}(p) & \Delta^{\bar{T}}(p) \end{pmatrix} = \begin{pmatrix} \dfrac{1}{p^2 - M^2 + \mathrm{i}\epsilon} & 0 \\ 0 & \dfrac{-1}{p^2 - M^2 - \mathrm{i}\epsilon} \end{pmatrix} + \begin{pmatrix} \tilde{c}_1(p)\delta(p^2 - M^2) & \tilde{c}_3(p)\delta(p^2 - M^2) \\ \tilde{c}_2(p)\delta(p^2 - M^2) & \tilde{c}_4(p)\delta(p^2 - M^2) \end{pmatrix}. \tag{4.11}$$

We will then argue the full form of these propagators using the constraints of unitarity, Hermiticity and CPT invariance [14] [1]. Let us begin with unitarity. It is straightforward to show that either using unitarity or the definitions of the scalar propagators given in equation (4.10) the following identity holds

$$\begin{aligned} i\Delta^{T}(x, y) + i\Delta^{\bar{T}}(x, y) &= i\Delta^{<}(x, y) + i\Delta^{>}(x, y) \\ i\Delta^{T}(p) + i\Delta^{\bar{T}}(p) &= i\Delta^{<}(p) + i\Delta^{>}(p). \end{aligned} \tag{4.12}$$

This, in combination with the identity

$$\frac{1}{p^2 - M^2 + i\epsilon} - \frac{1}{p^2 - M^2 - i\epsilon} = -2\pi i\delta\left(p^2 - M^2\right), \tag{4.13}$$

gives the relation

$$-2\pi i + (\tilde{c}_1 + \tilde{c}_4) = \tilde{c}_2 + \tilde{c}_3. \tag{4.14}$$

Then we can use CPT invariance and Hermiticity to further constrain these functions. To use CPT invariance we recall the effect of charge conjugation operator acting on a scalar field

$$\begin{aligned} U_C^\dagger \phi(x) U_C &= e^{i\eta}\phi^\dagger(x) \\ U_C^\dagger \phi^\dagger(x) U_C &= e^{-i\eta}\phi(x). \end{aligned} \tag{4.15}$$

We then obtain the following relationships between the coefficients in our generic form of our propagator

$$\begin{aligned} \Delta^{>}(p) &= -\Delta^{*>}(p) = \Delta^{C<}(-p) = -\Delta^{*C<}(-p) \\ \Delta^{T}(p) &== \Delta^{CT}(-p) = -\Delta^{C\bar{T}}(p) = -\Delta^{*C\bar{T}}. \end{aligned} \tag{4.16}$$

These relations give the relations

$$\begin{aligned} &\tilde{c}_1(p) = \tilde{c}_1(-p),\ \tilde{c}_4(p) = \tilde{c}_4(-p),\ \tilde{c}_2(p) = \tilde{c}_3(-p) \\ &\tilde{c}_1(p) = -\tilde{c}_4^*(p),\ \mathrm{Re}\,[\tilde{c}_2(p)] = \mathrm{Re}\,[\tilde{c}_3(p)] = 0. \end{aligned} \tag{4.17}$$

The most general solution to the above solutions consistent with isotropy and spatial homogeneity[2] and taking note of the fact that $a^\dagger a = n_B(|p_0|)\delta^3(p - k)$ is

$$\begin{pmatrix} \tilde{c}_1(p)\delta\left(p^2 - M^2\right) & \tilde{c}_3(p)\delta\left(p^2 - M^2\right) \\ \tilde{c}_2(p)\delta\left(p^2 - M^2\right) & \tilde{c}_4(p)\delta\left(p^2 - M^2\right) \end{pmatrix} \tag{4.18}$$

$$= -2\pi i\left[n_B\left(|p_0|\right)\begin{pmatrix} 1 & 1 \\ 1 & 1 \end{pmatrix} + \begin{pmatrix} 0 & \Theta(-p_0) \\ \Theta(p_0) & 0 \end{pmatrix}\right].$$

[1] Causality is also indirectly implied by the form of our propagators.

[2] We also imply causality though we have not explicitly discussed it here.

A similar derivation for fermions gives that the correction to the zero temperature propagators are given by

$$\begin{pmatrix} S^T(p) & S^<(p) \\ S^>(p) & S^{\bar{T}} \end{pmatrix} = \begin{pmatrix} \dfrac{\not{p} + m}{p^2 - m^2 + i\epsilon} & 0 \\ 0 & \dfrac{\not{p} + m}{p^2 - m^2 - i\epsilon} \end{pmatrix} \\ + 2\pi i(\not{p} + m)\left[n_F(|p_0|)\begin{pmatrix} 1 & 1 \\ 1 & 1 \end{pmatrix} + \begin{pmatrix} 0 & -\Theta(-p_0) \\ -\Theta(p_0) & 0 \end{pmatrix} \right]. \tag{4.19}$$

The propagators and the closed time path contour turn out to be all we need to understand to manipulate the effective potential at high temperature.

4.2 A brief review of the effective potential at zero temperature

The effective potential receives finite temperature corrections at loop level that can qualitatively change its minima and therefore the vacuum expectation value. This topic at zero temperature is generally treated in graduate level courses [5, 15, 27], however, diving straight into the finite temperature treatment of the effective potential can be disorientating. Therefore to warm up we will review how to calculate the one-loop corrections to a simple scalar field theory at zero temperature. We also omit contributions from fermions based on the fact that this topic is elementary and just including the scalar self-interactions is a compromise between not covering too much elementary review and easing the transition to finite temperature calculations. The goal of this short review is threefold—to demonstrate that the one-loop correction to the effective potential can be written as a geometric series in the propagator, to review what the forms of the one-loop corrections to the effective potential look like after regularization and renormalization, and to introduce a trick that simplifies the calculation by taking the derivative of the one-loop correction to the effective potential and integrating. To derive the effective potential we begin as usual from the generating functional

$$iW_i[j] \equiv \log z[j] = \log\left[\int \mathrm{d}\phi \exp(S[\phi] + \phi j \exp)\right]. \tag{4.20}$$

Here spatial integrals and arguments have been suppressed for notational convenience. As indicated by the argument, the natural variable of $W[j]$ is j. That is,

$$\mathrm{d}W = \frac{\partial W}{\partial j}\mathrm{d}j = \phi \mathrm{d}j. \tag{4.21}$$

To change the natural variables of W from j to ϕ, one simply performs a Legendre transformation:

$$\Gamma[\phi] = W[j] - \phi j. \tag{4.22}$$

The left-hand side of the above equation is the effective action. The effective potential for a translational invariant theory is defined by calculating Γ for a constant field configuration

$$\Gamma(\phi_c) = -\int \mathrm{d}^4x\, V_{\mathrm{eff}}(\phi_c). \tag{4.23}$$

To calculate the effective potential, we can Taylor expand the effective action in powers of the field

$$\Gamma[\phi] = \sum_{n=0}^{\infty} \frac{\phi^n}{n!} \frac{\partial^n \Gamma(\phi)}{\partial \phi}\bigg|_{\phi=0}. \tag{4.24}$$

Let us now restore the spatial dependencies and integrals

$$\Gamma[\phi] = \sum_{n=0}^{\infty} \frac{1}{n!} \int \prod_{i=1}^{n} \mathrm{d}^4x_i \phi(x_i) \Gamma^n(x_1 \cdots x_n). \tag{4.25}$$

Here $\Gamma^n(x)$ are the one-particle irreducible Green's functions. For a scalar field theory with quartic interactions these functions are given in figure 4.4. Let us take the Fourier transform of these Green's functions.

$$\Gamma^n(x) = \int \left(\prod_{i=1}^{n} \frac{\mathrm{d}^4 p_i}{(2\pi)^4} \exp(\mathrm{i} p_i x_i) \right) (2\pi)^4 \delta^4(p_1 \cdots + p_n) \Gamma^n(p), \tag{4.26}$$

where

$$\phi(p) = \int \mathrm{d}^x \mathrm{e}^{-\mathrm{i}px} \phi(x). \tag{4.27}$$

So we have as our expression for the effective action

$$\Gamma(\phi_c) = \frac{1}{n!} \sum_{n=0}^{\infty} \int \prod_{i=1}^{n} \left[\frac{\mathrm{d}^4 p_i \phi(-p_i)}{(2\pi)^4} \right] (2\pi)^4 \delta^{(4)}(p_1 + \cdots + p_n) \Gamma^n(p_1 + \cdots + p_n) \bigg|_{\phi_c}. \tag{4.28}$$

For a simple example that gives us most of the essential physics consider a model with a single scalar field whose tree-level potential is given by

$$V_0 = \frac{1}{2} m^2 \phi^2 + \frac{\lambda}{4!} \phi^4. \tag{4.29}$$

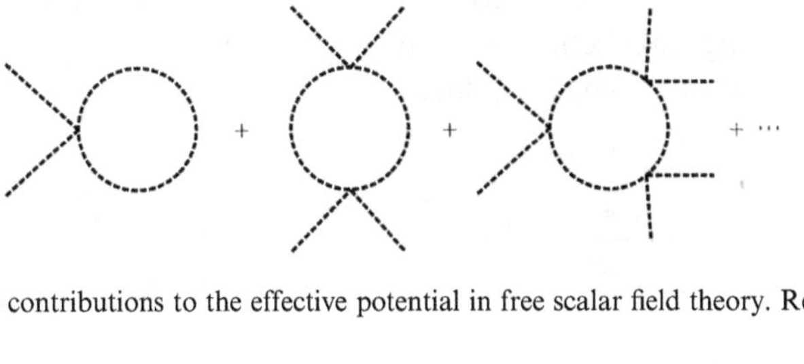

Figure 4.4. One-loop contributions to the effective potential in free scalar field theory. Reproduced from [5].

Our goal is to calculate the effective potential to one loop. To do so we will set the scalar field to be a constant in space–time as stated before. In momentum space the scalar field is therefore

$$\phi_c(p) = (2\pi)^4 \phi_c \delta^4(p). \tag{4.30}$$

Substituting this into the expression we have for the effective potential gives

$$\Gamma[\phi_c] = -\sum_{n=0}^{\infty} \frac{\phi_c^n}{n!} \Gamma^n(p_i = 0). \tag{4.31}$$

Using the explicit form for one particle irreducible diagrams that are at most one loop from figure 4.4 we see that the above sum is actually a geometric series

$$\begin{aligned} V_1 &= i \sum_{n=1}^{\infty} \int \frac{\mathrm{d}^4 p}{(2\pi)^4} \frac{1}{2n} \left[\frac{\lambda \phi_c^2 / 2}{p^2 - m^2 + i\epsilon} \right]^n \\ &= \frac{1}{2} \int \frac{\mathrm{d}^4 p}{(2\pi)^4} \log\left[p^2 + m^2(\phi) \right]. \end{aligned} \tag{4.32}$$

We can calculate this easily using a hard cutoff regulator

$$\begin{aligned} V_1 &= \frac{1}{32\pi^2} \int_0^{\Lambda^2} p^2 \log\left[p^2 + m^2 \right] \mathrm{d}p^2 \\ &= \frac{1}{32\pi^2} \left[\frac{1}{2}(p^4 - m^4)\log(p^2 + m^2) - \frac{1}{2}\left(\frac{p^4}{2} - p^2 m^2 \right) \right]_0^{\Lambda^2}. \end{aligned} \tag{4.33}$$

So in total we have the one-loop correction to the effective potential at zero temperature

$$V_1 = \frac{1}{32\pi^2} \left[\frac{1}{2}\left(\Lambda^4 - m^4 \right)\log\left(\Lambda^2 + m^2 \right) - \frac{1}{2}\left(\frac{\Lambda^4}{2} - \Lambda^2 m^2 \right) + \frac{m^4}{2} \log(m^2) \right]. \tag{4.34}$$

We can expand the first log term into two logarithms $\log(\Lambda^2 + m^2) = \log(\Lambda^2) + \log(1 + m^2/\Lambda^2)$. The second logarithm can be expanded to second-order. Taking the limit $\Lambda \to \infty$ and ignoring m-independent terms we can write

$$V = \frac{1}{32\pi^2} m^2 \Lambda^2 + \frac{1}{64\pi^2} m^4 \left[\log \frac{m^2}{\Lambda^2} - \frac{1}{2} \right]. \tag{4.35}$$

Finally we want to revise how to renormalize the one-loop zero temperature effective action using two schemes of regularization—finite cutoff regularization as used above and dimensional regularization. Let us introduce counter terms into our Lagrangian

$$V = \frac{1}{2} \delta m^2 \phi_c^2 + \frac{\lambda + \delta\lambda}{4!} \phi_c^4 + \frac{\lambda \phi_c^2 \Lambda^2}{64\pi^2} + \frac{\lambda^2 \phi_c^4}{256\pi^2} \left(\log \frac{\lambda \phi_c^2}{\Lambda^2} - \frac{1}{2} \right). \tag{4.36}$$

We are free to choose a renormalization scheme where

$$\left.\frac{\mathrm{d}^2 V}{\mathrm{d}\phi^2}\right|_{\phi=0} = 0, \quad \left.\frac{\mathrm{d}^4 V}{\mathrm{d}\phi^4}\right|_{\phi=\mu} = \lambda \tag{4.37}$$

which gives the conditions

$$\delta m^2 = \frac{\lambda\Lambda^2}{64\pi^2} \tag{4.38}$$

$$\delta\lambda = -\frac{\lambda^2}{32\pi^2} - \frac{3\lambda^2}{32\pi^2}\log\frac{\lambda\mu^2}{2\Lambda^2}. \tag{4.39}$$

From this we obtain the renormalized effective potential using a cutoff regularization scheme

$$V_{\text{eff}} = \frac{\lambda\phi^4}{4!} + \frac{\lambda^2\phi^4}{256\pi^2}\left(\log\frac{\phi^2}{\mu^2} - \frac{25}{6}\right). \tag{4.40}$$

We will conclude by calculating the renormalized effective potential in dimensional regularization and showing a trick which simplifies the calculation considerably. Our one-loop correction to our potential for $d = 4 - \epsilon$ is

$$V_1(\phi_c) = \frac{1}{2}\left(\mu^2\right)^{2-\frac{n}{2}} \int \frac{\mathrm{d}^n p}{(2\pi)^n} \log\left[p^2 + m^2\right]. \tag{4.41}$$

It is much easier to calculate the one-loop correction to the effective potential by first calculating it with respect to the mass and then integrating. The derivative is just a single disconnected bubble

$$\begin{aligned} \frac{\partial V_1}{\partial m^2} &= \frac{1}{2}\left(\mu^2\right)^{2-\frac{n}{2}} \int \frac{\mathrm{d}^n p}{(2\pi)^n}\frac{1}{p^2 + m^2} \\ V_1 &= \frac{m^4}{64\pi}\left(-\left[\frac{1}{\epsilon} - \gamma_E + \log 4\pi\right] + \log\frac{m^2}{\mu^2} - \frac{3}{2}\right), \end{aligned} \tag{4.42}$$

where we have used the relation $A^\epsilon \approx 1 - \epsilon \log A$. We can subtract the $1/\epsilon - \gamma_E - \log 4\pi$ term to obtain

$$V_1 = \frac{1}{64\pi^2}m^4\left(\log\frac{m^2}{\mu^2} - \frac{3}{2}\right), \tag{4.43}$$

where m^2 is the tree-level field-dependent mass defined by $m^2 = \partial^2 V_0(\phi)$. Here we have used a mass-independent renormalization scale μ. The perturbative expansion is most accurate when μ is about the scale of the maximum value of $m_i(\phi)$.

4.3 The effective potential at finite temperature

From the above revision of the zero temperature effective potential one can guess that the derivation of the one-loop effective potential at finite temperature involves summing over all one-particle irreducible interactions at finite temperature. This can be quite cumbersome—particularly in the closed time path formalism—as there are now four Green's functions instead of one. However, a dramatic simplification can be made by using the trick at the end of the previous subsection where we observed that the derivative of one-loop corrections to shifted effective potentials, $\partial V/\partial m^2(\phi_c)$, are diagrammaticaly equivalent to disconnected bubble diagrams. One can then calculate this diagram(s) and integrate. Returning to the simple example of ϕ^4 theory, we have only one bubble diagram to compute and in fact only one propagator to concern ourselves with, since external legs must be on the positive time contour (that is the contour that goes from $-\infty$ to the present ϵ above the real axis):

$$\frac{\partial V}{\partial m^2(\phi_c)} = \frac{1}{2}\int \frac{\mathrm{d}^4 p}{(2\pi)^4}\left[\frac{\mathrm{i}}{p^2 - m^2(\phi) + \mathrm{i}\epsilon} + 2\pi n_B(\omega_p)\delta(p^2 - m^2)\right]. \tag{4.44}$$

So we see the one-loop correction is a sum of the zero temperature component and the finite temperature component. Let us focus exclusively on the finite temperature component. The time integral can be calculated by using the identity

$$\delta(p^2 - m^2) = \frac{1}{2\omega_p}\left[\delta(p_0 + \omega_o) + \delta(p_0 - \omega_p)\right], \tag{4.45}$$

which allows us to write the finite temperature part of the bubble

$$\frac{\partial V_1(T)}{\partial m^2(\phi_c)} = \int \frac{\mathrm{d}^3 p}{(2\pi)^3}\frac{1}{2\omega} n_B(\omega). \tag{4.46}$$

Let us make the convenient substitution $p = Tx$. It is easy to show by direct substitution that the integral of the above equation is [16]

$$V_1(T) = \frac{T^4}{2\pi^2} J_B\left(\frac{m^2(\phi_x)}{T^2}\right) \equiv \frac{T^4}{2\pi^2}\int_0^\infty \mathrm{d}x x^2 \log\left[1 - \mathrm{e}^{-\sqrt{x^2 + \frac{m^2(\phi_c)}{T^2}}}\right]. \tag{4.47}$$

This is just the free energy of a relativistic gas of bosons. Next we turn our attention to fermion contributions to the one-loop finite temperature effective potential. The process is identical to the above where one considers the derivative of the effective potential, $\partial V_1/\partial m_f(\phi_c)$, and integrates. In this case we obtain the expression

$$V_1(T) = \frac{T^4}{2\pi^2} J_F\left(\frac{m_f^2(\phi_x)}{T^2}\right) \equiv \frac{T^4}{2\pi^2}\int_0^\infty \mathrm{d}x x^2 \log\left[1 + \mathrm{e}^{-\sqrt{x^2 + \frac{m_f^2(\phi_c)}{T^2}}}\right], \tag{4.48}$$

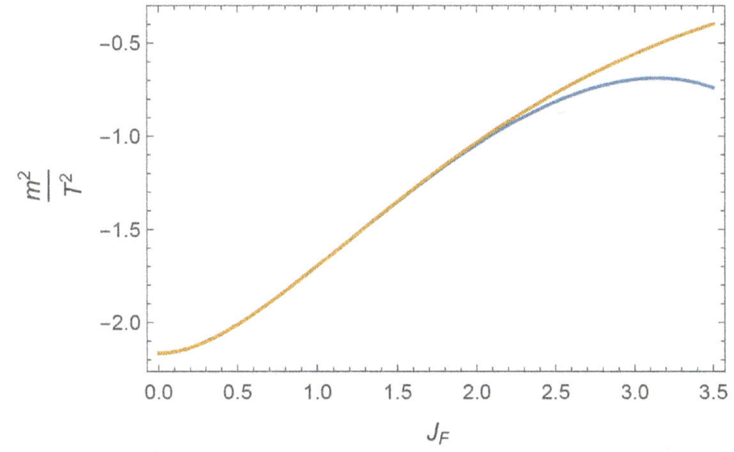

Figure 4.5. $J_B(m^2/T^2)$ compared to a high temperature expansion extended to terms of $O(m^8/T^8)$. The golden line again is the full thermal function, whereas the blue line is the high temperature expansion. The higher order terms make only a slight extension to the region of validity. Note the Riemann zeta function $\zeta(5)$ in equation (4.51) has been approximated to be 1. Fine-tuning this value makes little difference.

which is the free energy of a relativistic gas of fermions. In practice, these integral expressions can be cumbersome to work with. It is useful to make an expansion into a polynomial in the fields. There is a high temperature expansion for each case which is valid for relatively small m^2/T^2. For the bosonic thermal function the expansion goes as

$$J_B(z^2) \approx -\frac{\pi^4}{45} + \frac{\pi^2 z^2}{12} - \frac{\pi z^3}{6} - \frac{z^4}{32}\left(\log z^2 - 5.4076\right) + \Delta J_B\left(z^2\right). \tag{4.49}$$

As seen in figure 4.2, the high temperature expansion works quite well for arguments up to ~2. The higher order corrections

$$\Delta J_B\left(z^2\right) = -2\pi^{7/2}\sum_{l=1}^{\infty}(-1)^l\frac{\zeta(2l+1)}{(l+1)!}\Gamma\left(l+\frac{1}{2}\right)\left(\frac{z^2}{4\pi^2}\right)^{l+2} \tag{4.50}$$

extend the region of validity somewhat. This we show in figure 4.5, where the first two terms in this series moderately affect the region of validity. The high temperature fermion expansion

$$J_F\left(z^2\right) \approx \frac{7\pi^4}{360} - \frac{\pi^2 z^2}{24} - \frac{z^4}{32}\left(\log z^2 - 2.6351\right) + \Delta J_F\left(z^2\right) \tag{4.51}$$

similarly works fairly well for arguments up to about ~2, as seen in figure 4.6, but unfortunately the higher order terms

$$\Delta J_F\left(z^2\right) = -\frac{\pi^{7/2}}{4}\sum_{l=1}^{\infty}(-1)^l\frac{\zeta(2l+1}{(l+1)!}\left(1-2^{-2l-1}\right)\Gamma\left(l+\frac{1}{2}\right)\left(\frac{z^2}{\pi^2}\right)^{l+2} \tag{4.52}$$

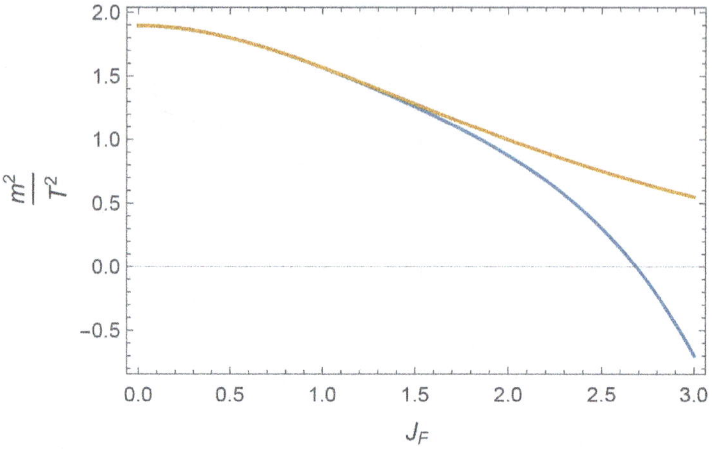

Figure 4.6. $J_F(m^2/T^2)$ compared to its high temperature expansion. The approximation works for $m^2/T^2 \lesssim 2$. The golden line is the full thermal function, whereas the blue line is the approximation.

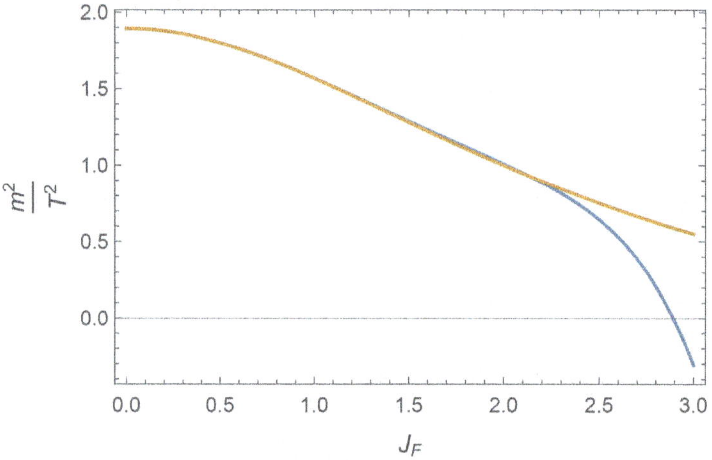

Figure 4.7. $J_F(m^2/T^2)$ compared to a high temperature expansion extended to terms of $O(m^8/T^8)$. The golden line is the full thermal function, whereas the blue line is the high temperature expansion. The higher order terms make only a slight extension to the region of validity. Note the Riemann zeta function $\zeta(5)$ in equation (4.53) has been approximated to be 1. Fine-tuning this value makes little difference.

suffer from the fact that they converge slowly, as shown in figure 4.7, where the first two correction terms barely extend the region of validity. When the field-dependent masses are very large compared to the temperature, the thermal functions contain Boltzman suppression,

$$J_B(z^2) = J_F(z^2) \approx \left(\frac{z}{2\pi}\right)^{\frac{3}{2}} e^{-z}\left[1 + O(z^{-1})\right], \tag{4.53}$$

so we can safely ignore the contributions to the electroweak phase transition for very heavy particles.

4.3.1 The Standard Model example

The Standard Model is a simple enough regime that within it you can treat the electroweak phase transition analytically within the high temperature expansion, given a couple of approximations. The Standard Model Higgs is a scalar, complex, SU(2) doublet so it has four degrees of freedom which can be put into a column vector h_i. Let us first approximate the full finite temperature effective potential by only considering contributions from the massive gauge bosons, Goldstone bosons, and the top quark as these contributions dominate. We will work in the Feynman gauge, $\xi = 0$ for simplicity, returning to the issue of gauge invariance later. The mass expansion in this case is neat since these masses are all proportional to the vacuum expectation value. We also include the quadratic thermal correction to μ including contributions from Goldstone bosons. If we are loose with concerns of gauge invariance (an issue we return to in a later section) we can write

$$V(H, T) \approx \left(-\frac{\mu^2}{2} + c_h T^2\right)h^2 - ETh^3 + \frac{\lambda}{4}h^4 \tag{4.54}$$

for coefficients

$$\begin{aligned} c_h &= \frac{1}{32}\left(g_1^2 + 3g_2^2 + 4y_t^2 + 8\lambda\right) \\ E &= \frac{3}{96\pi}\left(2g_2^3 + \left(g_1^2 + g_2^2\right)^{3/2}\right). \end{aligned} \tag{4.55}$$

Qualitatively one can describe from the form of the above equations how the Standard Model effective potential evolves with temperature. At some certain temperature $c_h T^2 > \mu^2$ and the coefficient of the bilinear term in the effective potential changes sign. The cubic term is negative definite, and if the parameters of the Standard Model occupy a particular range then the potential will have an additional turning point appear which is a local maximum between the origin and the global minimum. Outside such a range the minimum will evolve ever closer to the origin until it disappears. Since the bilinear term in the potential is quadratic in the temperature, whereas the cubic grows linearly with temperature, the third turning point will eventually disappear at very high temperature and the potential will have a single minimum at the origin.

Recall a first-order phase transition is characterized by a bump separating the origin and the global minimum whereas a second-order phase transition is characterized by the new minimum just rolling away from the origin continuously as the system cools. At very high temperature the potential has a single minimum at the origin with no bumps or wiggles. At a lower temperature the minimum becomes degenerate. This is known as the critical temperature. At a much lower temperature the barrier disappears at which point the phase transition is well and truly complete. We can define the conditions for the critical temperature as

$$V(h_C, T_C) = 0$$
$$\left.\frac{\partial V(h, T)}{\partial h}\right|_{h_C, T_C} = 0, \tag{4.56}$$

which is easy to solve in the Standard Model. A simple calculation gives

$$h_C = \frac{E}{\sqrt{c_h}} \frac{\mu^2}{\sqrt{\lambda\left(c_h\lambda - E^2\right)}}$$
$$T_C = \frac{\lambda}{2\sqrt{c_h}} \frac{\mu^2}{\sqrt{\lambda\left(c_h\lambda - E^2\right)}}, \tag{4.57}$$

which gives the simple relationship for the strength of the phase transition that can be written in terms of either the quartic coupling or the tree-level $0T$ Higgs mass

$$\frac{h_c}{T_C} = \frac{E}{2\lambda} = \frac{E v_H^2}{m_H^2}. \tag{4.58}$$

For gauge couplings $g_1 \sim 0.35$, $g_2 \sim 0.65$ we have $E = 0.0067$ which means we need the tree-level Higgs mass to be about a twelfth of the vacuum expectation value to obtain an order parameter of unity. Clearly the Higgs mass is far too heavy to accommodate this. If the electroweak phase transition is strongly first-order we need extra bosons to couple to the Higgs.

4.3.2 Issues of gauge invariance

So far our analysis of the Standard Model effective potential completely ignores gauge invariance. Gauge dependence manifests itself at the one-loop level since the propagators are gauge-dependent [17]. The Higgs field has four degrees of freedom which we can put into a column vector $\phi = (\phi_1, \phi_2, \phi_3, \phi_4)$ along with the generating function $j = j_i$ to simplify notation. At this stage to make the analysis as general as possible we will consider a model with a scalar field that couples to a non-abelian gauge field. The gauge fixed generating functional for such a model is

$$Z[j] = \int \mathcal{D}\phi\mathcal{D}A\mathcal{D}\eta\mathcal{D}\eta^\dagger e^{i\int d^4x(\mathcal{L}(x, j, \xi))} \tag{4.59}$$

with

$$
\begin{aligned}
\mathcal{L}(x, j, \xi) = &- V(\phi_{\mathrm{cl}}) + j\phi_{\mathrm{cl}} + \phi_i\left(-\frac{\partial V}{\partial \phi} + j_i\right) \\
&+ \frac{1}{2}\phi_i\left[-\partial^2 - M_{ij}^2(\phi_{\mathrm{cl}}) - \xi m_A^2(\phi_{\mathrm{cl}})_{ij}\right]\phi_j \\
&+ \frac{1}{2}A_\mu^a\left[\left(\partial^2 g^{\mu\nu} - \left(1 - \frac{1}{\xi}\partial^\mu\partial^\nu\right)\delta^{ab} + m_A^2(\phi_{\mathrm{cl}})^{ab} g^{\mu\nu}\right]A_\nu^a \\
&+ \eta^{\dagger a}\left[-\partial^2\delta^{ab} - \xi m_A^2(\phi_{\mathrm{cl}})^{ab}\right]\eta^b
\end{aligned}
\tag{4.60}
$$

and the field-dependent masses are defined

$$
\begin{aligned}
M_{ij}^2(\phi_{\mathrm{cl}}) &= \left.\frac{\partial^2 V}{\partial\phi_i\partial\phi_j}\right|_{\phi_{\mathrm{cl}}} \\
m_A^2(\phi_{\mathrm{cl}})^{ab} &= (gT^a\phi_{\mathrm{cl}})_i(gT^b\phi_{\mathrm{cl}})_j \\
\xi m_A^2(\phi_{\mathrm{cl}})_{ij} &= \xi m_A^2(\phi_{\mathrm{cl}})^{aa}.
\end{aligned}
\tag{4.61}
$$

Following the same steps as outlined in the previous sections, one finds that the effective potential to one loop, ignoring fermions, is

$$
\begin{aligned}
V = V_{\mathrm{tree}}(\phi) &+ \sum_{\mathrm{scalars}, i}\frac{\left[m_i^2(\phi, \xi)\right]^2}{64\pi^2}\left[\log\left(\frac{m_i^2(\phi, \xi)}{\mu^2}\right) - \frac{3}{2}\right] \\
&+ 3\sum_{\mathrm{gauge}, a}\frac{\left[m_a^2(\phi)\right]^2}{64\pi^2}\left[\log\left(\frac{m_a^2(\phi)}{\mu^2}\right) - \frac{5}{6}\right] \\
&- \sum_{\mathrm{gauge}, a}\frac{\left[\xi m_a^2(\phi)\right]^2}{64\pi^2}\left[\log\left(\frac{\xi m_a^2(\phi)}{\mu^2}\right) - \frac{3}{2}\right] \\
&+ \frac{T^4}{2\pi^2}\left[\sum_{\mathrm{scalar}, i} J_B\left(\frac{m_i^2(\phi, \xi)}{T^2}\right) + 3\sum_{\mathrm{gauge}, a} J_B\left(\frac{m_a^2(\phi)}{T^2}\right) - \sum_{\mathrm{gauge}, a} J_B\left(\frac{\xi m_a^2(\phi)}{T^2}\right)\right].
\end{aligned}
\tag{4.62}
$$

Making the same approximations as before one finds that the order parameter is gauge-dependent

$$
\frac{\phi_C}{T_C} = \frac{2E}{\lambda}\left(1 - \frac{1}{3}\xi^{3/2}\right). \tag{4.63}
$$

In fact a particular choice of gauge, $\xi = 3^{2/3}$ sets the order parameter to zero regardless of the value of E or λ! All physical properties have to be gauge-independent by definition and the order parameter is indeed a physical quantity—the point where any reduction in temperature makes it energetically favorable to tunnel from the false vacuum to the true vacuum. So if we find the order parameter has become gauge-dependent then it must be an artifact of our approximation scheme in expanding the effective action at finite temperature to one loop and finding the minima.

In order to calculate a gauge-independent order parameter we therefore need to find a gauge-independent expansion of v_c and T_C or $V_{\rm eff}(T)$. Making an expansion in the temperature turns out to be problematic at the two-loop level so we will follow the prescription set out in [17]. To do this we make use of the Nielsen identity [18, 19],

$$\frac{\partial V_{\rm eff}}{\partial \xi} = -C_i(\phi, \xi)\frac{\partial V_{\rm eff}}{\partial \phi_i}, \tag{4.64}$$

which means the effective potential is gauge-independent at its extrema. However, the extrema at the one-loop level are gauge-dependent. Since the extrema of the full effective potential must be gauge-independent (as it is a physical quantity) again the gauge dependence of the minimum must be an artifact of our approximation scheme.

Let us expand both C_i and the effective potential as a series in $\hbar$. That is $V_{\rm eff} = V_0(\phi) + \hbar V_1(\phi) + \cdots$ and $C = c_0 + \hbar c_1 + \cdots$, and V_1 is of course the one-loop correction to the effective potential. Substituting this into the Nielsen identity we obtain

$$\frac{\partial V_0}{\partial \xi} + \hbar\frac{\partial V_1}{\partial \xi} = -c_0\frac{\partial V_0}{\partial \phi} - \hbar\left(c_0\frac{\partial V_1}{\partial \phi} + c_1\frac{\partial V_0}{\partial \phi}\right). \tag{4.65}$$

We then equate terms of each power of $\hbar$ on either side to each other. The zeroth-order terms are straightforward to handle since the tree-level effective potential is manifestly gauge invariant which implies $c_0 = 0$. For terms linear in $\hbar$ we obtain

$$\frac{\partial V_1}{\partial \xi} = -c_1\frac{\partial V_0}{\partial \phi}, \tag{4.66}$$

which implies that the one-loop correction to the effective potential is gauge invariant at the tree-level minima, and *not* the one-loop minima. Let us make a $\hbar$ expansion to the minima as previously promised, $\phi_c = \phi_0 + \hbar\phi_1 + \cdots$, where ϕ_0 are the tree-level minima. It is then straightforward to find gauge invariant corrections to the tree-level minima as follows

$$\begin{aligned}\left.\frac{\partial V_{\rm eff}}{\partial \phi}\right|_{\phi_c} = 0 = \left.\frac{\partial V_0}{\partial \phi}\right|_{\phi_0+\hbar\phi_1+\cdots} + \hbar\left.\frac{\partial V_1}{\partial_\phi}\right|_{\phi_0+\hbar\phi+\cdots} \\ = \left.\frac{\partial V_0}{\partial \phi}\right|_{\phi_0} + \left.\left[\frac{\partial V_0}{\partial \phi} + \hbar\left(\frac{\partial V_1}{\partial \phi} + \phi_1\frac{\partial^2 V_0}{\partial \phi^2}\right)\right]\right|_{\phi=\phi_0} + \cdots.\end{aligned} \tag{4.67}$$

Insisting that each power of $\hbar$ vanishes gives a gauge invariant expression for ϕ_1

$$\phi_1 = -\left[\left(\frac{\partial^2 V_0}{\partial \phi^2}\right)^{-1} \frac{\partial V_1}{\partial \phi}\right]\Bigg|_{\phi=\phi_0}. \tag{4.68}$$

Substituting this back into our $\hbar$ expansion for the effective potential we can state a condition for the critical temperature which is valid to two loops

$$\begin{aligned} V(\phi_c, T_C) &= V_0(\phi_0) + \hbar V_1(\phi_0, T_C) \\ &+ \hbar^2\left[V_2(\phi_0, T_C) - \frac{1}{2}\phi_1(\phi_0)^2 \frac{\partial V_0}{\partial \phi}\bigg|_{\phi_0}\right] + \cdots = 0. \end{aligned} \tag{4.69}$$

The critical temperature is then evaluated by solving the above equation.

4.3.3 Daisy re-summation

At finite temperature perturbative theory becomes an expansion in the coupling constant multiplied by T^2/M^2. At some high enough temperature the theory will therefore become non-perturbative. In fact a theory can become non-perturbative near the critical temperature [20, 21] and this is indeed the case in the Standard Model if the Higgs mass is larger than the W boson mass. To deal with this we need to re-sum the dominant thermal corrections. This process is called 'daisy re-summation' and the result of the re-summation is that the field-dependent masses are given a thermal correction due to the one-loop self-energy. Daisy re-summation results in the generation of an additional term in the effective potential of the form

$$V_1^{\text{daisy}}(\phi, T) = -\frac{T}{12\pi}\sum n\left[\bar{m}^3(\phi, T) - m^3(\phi)\right]^{3/2}, \tag{4.70}$$

where $\bar{m}^2(\phi)$ is the field-dependent mass with a thermal correction

$$\bar{m}^2(\phi) = m^2(\phi) + \Delta_T m^2, \tag{4.71}$$

with the thermal mass correction having the structure of being a function of coupling constants times T^2. We show how to calculate thermal masses in section 7.4 as well as presenting the thermal masses from the MSSM. We return to the issue of re-summation there.

4.3.4 The sphaleron rate at high temperature

Now that we have the necessary knowledge of both finite temperature QFT and the evolution of the vacuum expectation value with temperature, we can fill in the gaps we previously left in the calculation of the sphaleron rate at high temperature [22, 23]. If one assumes that the sphaleron gas is very dilute we can expand around the single sphaleron solution. We will take the sphaleron rate to be the probability of finding a right moving particle at the barrier ($x = 0$) of the periodic potential times the rate at which it crosses the barrier. If the temperature is very large then it is justified to work in a classical regime where we simultaneously define position and momentum. This means $\Gamma = \langle\delta(x)\Theta(p)\rangle$ as the step function ensures a right-moving particle at a barrier which is at the origin. The classical rate is

$$\Gamma = \frac{\int \mathrm{d}p\mathrm{d}x \exp\left[-\beta\left(\frac{1}{2}p^2\right) + V(x)\right]\delta(x)\Theta(p)}{\int \mathrm{d}p\mathrm{d}x \exp\left[-\beta\left(\frac{1}{2}p^2\right) + V(x)\right]} \approx \frac{\omega_0}{2\pi}\mathrm{e}^{-\beta V_0}. \tag{4.72}$$

We can relate this to the imaginary part of the free energy of a single sphaleron with small fluctuations

$$\begin{aligned} \mathrm{Im}\, F &= T\, \mathrm{Im}\, \ln Z = T\frac{\mathrm{Im}\, Z}{Z} \\ &\approx T\frac{\mathrm{Im}\int \mathrm{d}p\mathrm{d}x \exp\left[-\beta\left(\frac{p^2}{2} + V_0 - \frac{1}{2}\omega x^2\right)\right]}{\int \mathrm{d}p\mathrm{d}x \exp\left[-\beta\left(\frac{p^2}{2} + \frac{1}{2}\omega x^2\right)\right]} \\ &= \frac{\omega_0}{2\omega_{-}\beta}\mathrm{e}^{-\beta V_0}, \end{aligned} \tag{4.73}$$

which means we can write

$$\Gamma \approx \frac{\omega_\beta}{\pi}\frac{\mathrm{Im}\, Z_{\mathrm{sph}}}{Z_0}. \tag{4.74}$$

Zero modes can be integrated out and each give a factor of $\mathcal{NV}$ where $\mathcal{V}$ is the volume of the symmetry group and $\mathcal{N}$ is a normalization factor. The sphaleron solution derived before has a $SU(2)_L$ spherical and translational symmetry so we obtain a factor of $(\mathcal{NV})_{\mathrm{tr}}$ as well as a factor of $(\mathcal{NV})_{\mathrm{rot}}$. Defining the coupling constant of the finite temperature electroweak theory as $g_3 = (g^2T/2M_W)^{1/2}$, the sphaleron rate is

$$\begin{aligned} \Gamma &\approx \frac{\omega_-}{2\pi}(\mathcal{NV})_{\mathrm{rot}}(\mathcal{NV})_{\mathrm{tr}}\, \mathrm{Im}\left[\frac{\det g_3^{-2}\Omega_0^2}{\det' g_3^{-2}\Omega_{\mathrm{sph}}^2}\right]^{1/2}\mathrm{e}^{-E_{\mathrm{sph}}/T} \\ &= \frac{\omega_-}{2\pi}(\mathcal{NV})_{\mathrm{rot}}(\mathcal{NV})_{\mathrm{tr}}\, \mathrm{Im}\left[\frac{\det \Omega_0^2}{\det' \Omega_{\mathrm{sph}}^2}\right]^{1/2} g_1^{-N_0}\mathrm{e}^{-E_{\mathrm{sph}}/T} \\ &\equiv \frac{\omega_-}{2\pi}(\mathcal{NV})_{\mathrm{rot}}(\mathcal{NV})_{\mathrm{tr}}\kappa g_3^{-N_0}\mathrm{e}^{-E_{\mathrm{sph}}/T} \\ &= \frac{\omega_-}{2\pi}(\mathcal{NV})_{\mathrm{rot}}(\mathcal{NV})_{\mathrm{tr}}\kappa\left(\frac{v(T)}{gT}\right)^3\mathrm{e}^{-E_{\mathrm{sph}}/T}. \end{aligned} \tag{4.75}$$

In the above we have implicitly defined the fluctuation determinant as $\kappa \approx 20$ for notational simplicity. As we derived before the sphaleron energy is given by

$$S_3\left[\phi_{\rm sph}\right] \equiv \frac{E_{\rm sph}}{T} = \frac{4\pi v(T)}{gT} B\left(\frac{\lambda}{g^2}\right) \tag{4.76}$$

with

$$\begin{aligned} B\left(\frac{\lambda}{g^2}\right) = \int_0^\infty dr \Bigg[4\left(\frac{df}{dr}\right)^2 + \frac{8}{r^2} f^2(1-f)^2 \\ + \frac{r^2}{2}\left(\frac{dh}{dr}\right)^2 h^2(1-f)^2 + \frac{1}{4}\frac{\lambda}{g^2} r^2 \left(1-h^2\right)^2 \Bigg], \end{aligned} \tag{4.77}$$

where $r \equiv gv(T)|x|^2$. In the symmetric phase the sphaleron rate can be determined from dimensional analysis with the only relevant scale being the thermal mass of the gauge bosons gT^2. Specifically the sphaleron rate is

$$\frac{\Gamma_{\rm sph}}{V} = 6\kappa\alpha_W^5 T. \tag{4.78}$$

Note that if there is a discontinuous jump in $v(T)$ as happens in a strongly first-order phase transition, the sphaleron rate can be much smaller than the Hubble rate, $H \approx \sqrt{g^*}\, T^2/M_{\rm pl}$, inside a bubble of broken phase, while being fast outside such a bubble. In figure 4.8 we show a schematic of the sphaleron rate compared to the Hubble rate turning a strongly first-order phase transition.

4.4 The bounce solution

The transition from the high temperature minimum to the true vacuum occurs via tunneling processes. We would like to understand the field profile that describes this tunneling process. The action has a local extremum known as the bounce solution where the fields continuously vary from one vacuum to the other only just tunneling through the barrier separating the two minima [25].

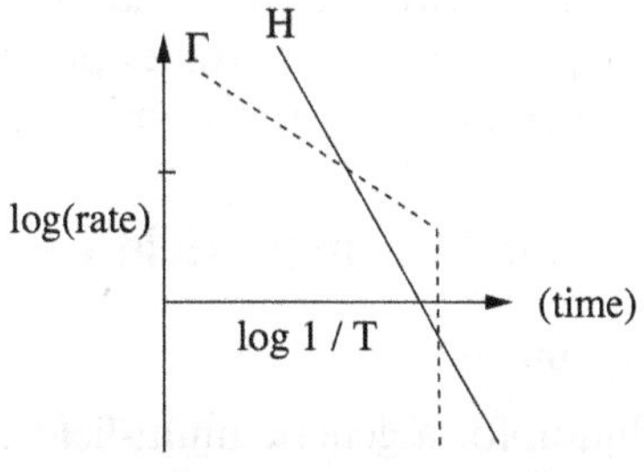

Figure 4.8. The sphaleron rate at high temperature in the advent of a strongly first-order phase transition. If the electroweak phase transition is sufficiently strongly first-order weak sphalerons drop out of equilibrium being fast compared to the Hubble rate outside the bubble and slow inside. Reproduced from [12].

Recall that the Euclidean action is

$$S_E = \int d^d x \left[\frac{1}{2} (\partial_\mu \phi_i)^2 + V(\phi_i) \right],$$

with $d = 3$ for finite temperature. The nucleation rate per unit volume is

$$\Gamma = AT^4 e^{-\frac{S_E}{T}}$$

[26, 27], where T is the temperature, A is the fluctuation determinant, and S_E is the Euclidean action for the bounce solution which satisfies the classical equations of motion. Numerically setting the expectation value for the number of bubbles per unit volume to unity, one finds $S_E/T \approx 140$ [28, 29] if the temperature is at the electroweak scale. More generally one finds $\frac{S_E}{T} = 170 - 4 \ln\left(\frac{T_N}{1\text{GeV}} \right)^{-2} \ln g_*$. Assuming a spherically symmetric bubble profile we can write the classical equations of motion in terms of a single space–time variable $\rho \equiv \sqrt{r^2 - t^2}$,

$$\frac{\partial^2 \phi_i}{\partial \rho^2} + \frac{(d-1)}{\rho} \frac{\partial \phi_i}{\partial \rho} - \frac{\partial V}{\partial \phi_i} = 0,$$

and the bounce solution satisfies the conditions $\phi_i(0) \approx v_i^{\text{true}}$, $\phi_i(\infty) = v_i^{\text{false}}$ and $\phi_i'(0) = 0^3$. Here ρ^2 is the space–time coordinate and v_i^{false} and v_i^{true} are the vacuum expectation values (VEVs) of the field ϕ_i in the true and false vacua, respectively. The classical equations of motion imitate a ball rolling in a landscape of shape $-V$ with ρ playing the role of time, but including a ρ-dependent friction term.

Using this analogy of a ball in a landscape, the true and false vacua are global and local maxima to the landscape, respectively, and they are separated by a valley. The solution that just pierces the barrier via tunneling processes is analogous to the ball starting from rest near the top of the global maximum and being released such that it comes to rest at $\rho = \infty$ with the ball settling on the local maximum. The time-dependent friction term means the ball loses energy along the journey so it must start at a point higher than the local maximum. If the starting point is too low it will come to rest at the bottom of the valley as $\rho \to \infty$. Alternatively if the starting point is too high the ball will go past the local maximum and diverge. The bounce solution is a fine-tuned point just between these two extremes and the total action (classically the action is approximately equal to energy times time) of the path that differs slightly from the bounce is clearly higher. So it is a local minimum.

4.5 Analytic techniques for the single field case

4.5.1 Single field approximation

Calculating the bounce solution for a generic multi-field case is a highly non-trivial task. Generally it is easier to approximate the system by a simpler system then use

[3] The first is not literally a boundary condition unlike the other two.

numerical or analytic techniques to improve the approximation [31, 32]. The simplified system that is generally used is a single-field problem.

To approximate the solution to the multi-field problem with a single-field solution, the steps are as follows. Begin by translating the effective potential such that the false vacuum is at the origin in field space. Then perform a rotation so that the true vacuum lies along one of the axes in the rotated frame in field space, say $\hat{\phi}_1$. Rewrite the effective potential in the rotated basis $V(\varphi_1, \varphi_2, \ldots) \to V(\phi_1, \phi_2, \ldots)$. Finally, one only considers the shape of the potential along the $\hat{\phi}_1$ direction by making the transformation

$$V(\phi_1, \phi_2, \ldots) \to V(\phi_1, 0, \ldots). \tag{4.79}$$

The equation of motion for the simplified potential is much more tractable

$$\frac{\partial^2 \phi_1}{\partial \rho^2} + \frac{(d-1)}{\rho}\frac{\partial \phi_1}{\partial \rho} - \frac{\partial V(\phi_1, 0, \ldots)}{\partial \phi_1} = 0.$$

The solution to this provides an ansatz that can be perturbed around to find the solution for the multi-field case.

Any system with three turning points where a local and global minimum are separated by a barrier can be approximated by the most general renormalizable potential with a single field

$$V(\varphi) = M^2\varphi^2 + b\varphi^3 + \lambda\varphi^4.$$

This includes the complicated curves that result from the complicated thermal functions[4]. Although it looks like there are three independent parameters, two can just be absorbed into rescaling transformations. For example, if we make the rescaling $\phi = \varphi_{\min}\varphi$, where $\varphi_{\min}$ is the global minimum, the rescaled potential will have its global minimum at $\phi = 1$. Such a rescaling needs to be compensated by a rescaling in the spatial parameters so that the effective action remains unchanged. This is achieved by the replacement $\rho \to \varphi_{\min}\rho$. A convenient parametrization for the rescaled potential is [34] [5]

$$V(\phi) = \frac{(4\alpha - 3)}{2}E\phi^2 + E\phi^3 - \alpha E\phi^4. \tag{4.80}$$

The strength of such a parametrization is that the second rescaling—stretching or contracting the V axis—is manifest by changing the value of E. Let us assume that the three turning points are in the positive ϕ direction (the other case is covered by reflection transformation). A barrier between the local and global maxima requires that $E < 0$ and $\alpha \in (0.5, 0.75)$. As shown in figure 4.9, increasing the value of α reduces the height of the barrier separating the minimum from the extreme cases of $\alpha = \frac{1}{2}$ where the minima are degenerate to $\alpha = \frac{3}{4}$ where the barrier disappears. Increasing $|E|$ increases the overall scale of the potential.

[4] An approximate expression for the effective action of a similar potential can be found in [33].

[5] This definition of α differs from that of [33] but the physical principles are the same.

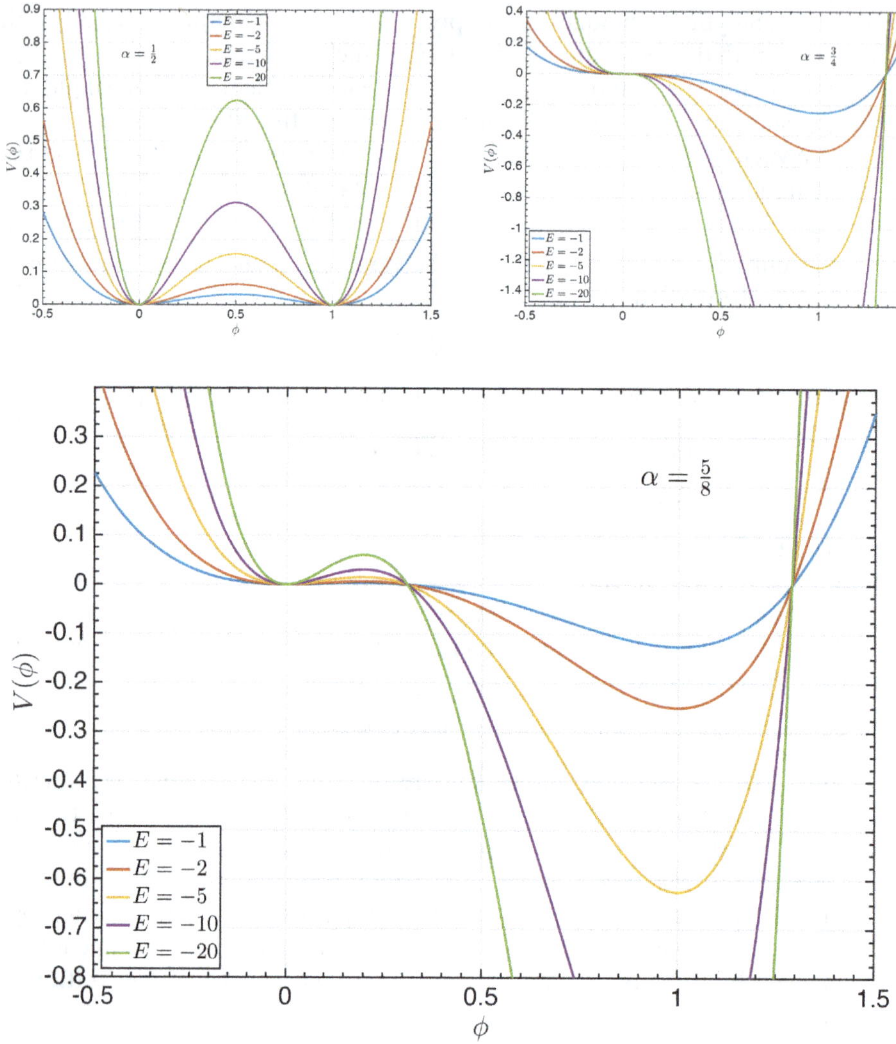

Figure 4.9. The most general tree-level renormalized potential rescaled such that the minimum is at one. The potential is parametrized by α and E which are given in equation (4.81). Every panel has the potential drawn for $-E = 1, 2, 5, 10, 20$. The parameter $\alpha \in [0.5, 0.75]$ is in a compact space. The larger panel has the mean value in $\alpha = \frac{5}{8}$ whereas the other two have the two edge cases with $\alpha = \frac{1}{2}$ on the left and $\alpha = \frac{3}{4}$ on the right.

4.5.2 Developing ansatz solutions

We can derive an approximate bounce solution to all potentials of the form given in equation (4.81) as it depends only on a single compact parameter, α. The parameter E is just a scale factor. Thus, one can factor $|E|$ out of the equations of motion by further rescaling $\rho \to \rho/\sqrt{|E|}$. Then, the equations of motion only depend on α.

Explicitly our rescaling operations are,

$$\begin{Bmatrix} \varphi \to \phi = \varphi_{\min}\varphi \\ \rho \to \dfrac{\varphi_{\min}}{\sqrt{|E|}}\rho \end{Bmatrix}$$

under which the Euclidean action becomes

$$S_E = 4\pi\frac{\phi_m^3}{\sqrt{|E|}}\int \mathrm{d}\rho\, \rho^2\left[\left(\frac{\partial\phi}{\partial\rho}\right)^2 - \tilde{V}(\phi)\right]$$

where[6] $\tilde{V} \equiv V/|E|$, and we have assumed isotropy which allows us to integrate over the solid angle. The effective action is now proportional to an integral that only depends on α, that is

$$S_E = 4\pi\frac{\phi_m^3}{\sqrt{|E|}}f(\alpha). \tag{4.81}$$

The bounces solutions can be found for the full parameter space via an undershoot overshoot method where one integrates the equation of motion numerically for various starting points until one finds the bounce as described in the previous section. These solutions can be approximated by kink solutions [35]

$$\phi \approx \frac{1}{2}\left(1 - \tanh\left[\frac{\rho - \delta(\alpha)}{L_w(\alpha)}\right]\right), \tag{4.82}$$

with parameters δ and L_w describing the offset and the bubble wall width respectively. If one curve fits the bounce solution to the kink solution numerically over the full parameter space, $\delta(\alpha)$ and $L_w(\alpha)$ become α-dependent functions. Also the integral in the Euclidean action, $f(\alpha)$, becomes α-dependent.

Some quick curve fitting over the entire parameter space gives the following functions for $\delta(\alpha)$, $L_w(\alpha)$ and $f(\alpha)$, respectively,

$$\delta(\alpha) \approx \delta_0 + k\left(\alpha - \frac{5}{8}\right) + \sum_{n=1}^{2} a_n\left[\frac{\alpha - \frac{5}{8}}{\left(\alpha - \frac{1}{2}\right)\left(\alpha - \frac{3}{4}\right)}\right]^{(2n-1)}$$

$$L_w(\alpha) \approx \ell_0\left[\left(\alpha - \frac{1}{2}\right)^r + \frac{c}{\left|\alpha - \frac{3}{4}\right|^s}\right]$$

[6] $\tilde{V}$ does not have any dependence on $|E|$.

Table 4.1. Table of fitter values that parametrize the functions that approximate the bubble wall width L_w, and the offset δ from the kink solution, as well as f which parametrizes the Euclidean action. Table reproduced from [34].

$f(\alpha)$		$L_w(\alpha)$		$\delta(\alpha)$	
Parameter	Value	Parameter	Value	Parameter	Value
f_0	0.0871	ℓ_0	1.4833	δ_0	2.2807
p	1.8335	c	0.4653	k	−4.6187
q	3.1416	r	18.0000	a_1	0.5211
		s	0.7035	$a_2 \times 10^5$	7.8756

and

$$f(\alpha) = f_0 \frac{\left|\alpha - \frac{3}{4}\right|^p}{\left|\alpha - \frac{1}{2}\right|^q}.$$

These functions are in excellent agreement with the numerical solution for the values given in table 4.1, which can be seen in figure 4.10.

Let us conclude with a note on how the tunneling rate depends on the remaining parameters of the single-field potential. First note that $|E|$ scales as $|b\phi^3|$. Therefore S_E/T scales as $\frac{\phi_m}{T}\sqrt{\frac{\phi_m}{b}}$, where b is the cubic coupling of the unscaled potential given in section 4.5.1 and therefore controls the height of the barrier.

4.6 Path deformation method

Any one-dimensional problem can be approximated fairly well by the set of kink solutions given in the previous section. One might ask if it is also possible in general to write an ansatz for the multi-field problem. Indeed such a method was attempted by [35] for a series of parametrized ansatzes

$$\phi_i(\rho) = \frac{\phi_{0i}}{2}\left(1 - \tanh\left[\frac{\rho + \delta_i}{L_i}\right]\right). \tag{4.83}$$

One can then in principle minimize the effective action with respect to the parameters $(\phi_{0i}, \delta_i, L_i)$. Unfortunately this method turns out to be virtually intractable numerically for a lot of potentials. The reason is that for particularly large values of $|\delta_i|$ or L_i the solution begins to resemble the solution of a constant field in a global minimum. As such some more sophisticated methods have been explored. We will give a detailed explanation of a semi-analytic method but first let us review a nice numerical method that involves deforming a path. This method takes advantage of the fact that you can always solve the one dimensional solution through bisection. Therefore, let us parametrize the path through field space which the tunneling

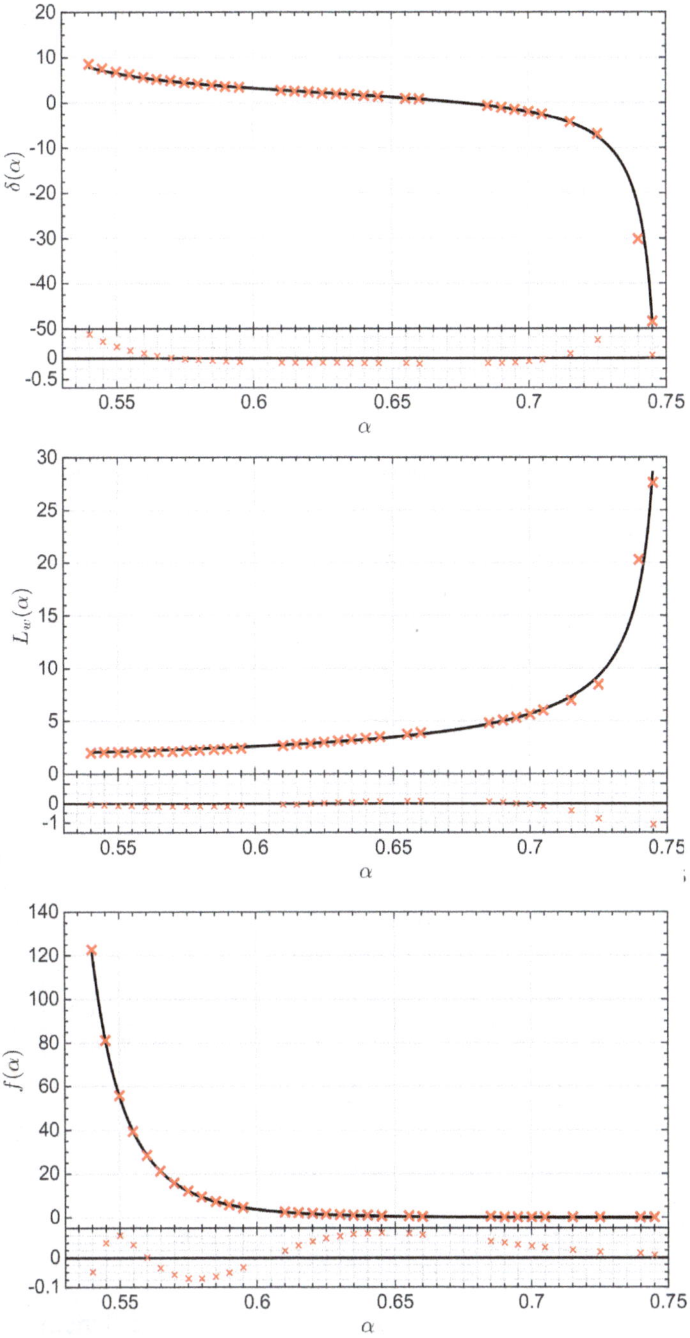

Figure 4.10. The fits to the bubble wall width L_w (top panel) and the offset δ (middle panel) of the kink solution, as well as the integral f (lowest), which appears in the Euclidean action. Both the numerically computed values and the residuals are given. Reproduced from [34].

solution follows by x. We can then separate the classical equations of motion into an equation along the path x and the orthogonal directions

$$\begin{aligned} \frac{d^2x}{d\rho^2} + \frac{2}{\rho}\frac{dx}{d\rho} &= \frac{\partial V}{\partial x} \\ \frac{\partial^2 \vec{\phi}}{dx^2}\left(\frac{dx}{d\rho}\right)^2 &= \nabla_\perp V. \end{aligned} \tag{4.84}$$

One can solve the first equation by bisecting the starting position x_0 along any given path and then deform the path by making steps in the orthogonal directions until the second set of equations is satisfied. Once the new path is defined one can solve the first equation along the new path. These two steps are repeated until convergence is achieved. One drawback to this method is that if the steps used to deform the path are too large then convergence might not be achieved. Nonetheless it is an elegant numerical solution that has been implemented in a publicly available package [32].

4.7 Perturbative method

Another method for finding the bubble wall profile can be found in [34] and involves finding an ansatz which is near to the true solution and perturbing around it. The key assumption is to assume that the trajectory in field space of the bounce solution does not differ from the single-field approximate trajectory by an amount that is greater than the distance between the two minima. In other words one assumes that

$$\left|\phi_i(\rho)\right| \lesssim \left|v_{\rm true} - v_{\rm false}\right|.$$

For the rescalings derived in section 4.5, the maximum distance between the ansatz and true solution is bounded by 1, although in practice it is usually much smaller. We will denote the ansatz solution for the field ϕ_i as A_i with perturbations denoted as ϵ_i. That is $\phi_i = A_i + \epsilon_i$. Inserting this into the classical equations of motion (4.85) yields

$$\frac{\partial^2 A_i}{\partial\rho^2} + \frac{\partial^2 \epsilon_i}{\partial\rho^2} + \frac{2}{\rho}\frac{\partial A_i}{\partial\rho} + \frac{2}{\rho}\frac{\partial \epsilon_i}{\partial\rho} = \left.\frac{\partial V(\phi)}{\partial\phi_i}\right|_A + \sum_j \left.\frac{\partial^2 V(\phi)}{\partial\phi_i\partial\phi_j}\right|_A \epsilon_j,$$

where $A \equiv \{A_i\}$. The above equation can be organized into terms involving the ansatz and terms involving the corrections. The result is a network of coupled inhomogeneous differential equations for the perturbations ϵ_i:

$$\frac{\partial^2 \epsilon_i}{\partial\rho^2} + \frac{2}{\rho}\frac{\partial \epsilon_i}{\partial\rho} - \left.\frac{\partial^2 V(\phi)}{\partial\phi_i\partial\phi_j}\right|_{A_j} \epsilon_i = B_i(\rho).$$

We have implicitly defined the inhomogeneous pieces of the differential equations as $B_i(\rho)$. These functions are given by

$$B_i(\rho) \equiv \left.\frac{\partial V(\phi)}{\partial\phi_i}\right|_A - \frac{\partial^2 A_i}{\partial\rho^2} - \frac{2}{\rho}\frac{\partial A_i}{\partial\rho}.$$

Clearly the function B_i is a measure of how well the ansatz solves the equations of motion, as one can see that if B_i is zero then the corrections to the ansatz are defined by a set of homogeneous differential equations that are satisfied by the trivial solution $\epsilon_i = 0$.

A key observation of [34] is that one can linearize and approximately solve these differential equations analytically by approximating the mass matrix by a series of step functions, with a correction which can be used to form the convergent series of perturbations. The techniques used to derive the solution are given in section 8.3. Here we just use the easier parametrized ansatz approach from that section where the form of the solution is assumed up to certain parameters and substituted back into the differential equations to determine such parameters.

To make use of the parametrized ansatz approach, consider the homogeneous form ($B_i = 0$) of the differential equations for ϵ. The solutions take the form

$$\epsilon \sim \frac{z e^{\lambda \rho}}{\rho}.$$

Inserting this into the set of differential equations results in an eigenvalue problem with the mass matrix. We parametrize the homogeneous solutions with the following notation

$$\epsilon_{ik}^{h} = \frac{z_{ik} e^{\lambda_k \rho}}{\rho}.$$

Here k spans the n eigenvalues of the mass matrix and is not summed over where as i is the field index. Substituting ϵ_{ik}^{h} for ϵ_i into the homogeneous form of the differential equations gives

$$\sum_j M_{ij} z_{jk} = \lambda_k^2 z_{ik},$$

whith M being the mass matrix

$$M_{ij} = \left. \frac{\partial^2 V(\phi)}{\partial \phi_i \partial \phi_j} \right|_A .$$

Note here that z_{ik} is the ith component for the eigenvector of the mass matrix with eigenvalue λ_k^2. Since the differential equations are second-order, one finds that ρ^2 are the eigenvalues of the mass matrix so the set of homogeneous solutions involve both the positive and negative roots, $\pm\lambda_k$. We therefore introduce $\tilde{\lambda}_j$ and $\tilde{z}_{ij}$, where

$$\tilde{\lambda}_1 = \lambda_1,\ \tilde{\lambda}_2 = -\lambda_1,\ \tilde{\lambda}_3 = \lambda_2,\ \ldots \tag{4.85}$$

$$\tilde{z}_{i1} = z_{i1},\ \tilde{z}_{i2} = z_{i1},\ \tilde{z}_{i3} = z_{i2},\ \tilde{z}_{i4} = z_{i2},\ \ldots \tag{4.86}$$

such that in $\tilde{\lambda}_j$ and $\tilde{z}_{ij}$, the index j spans the range 1, 2, …, $2n$. Using the techniques described in [36] one can derive that the particular solution is

$$\epsilon_i^{\lessgtr} = \sum_{j=1}^{2n}\sum_{k=1}^{n} \tilde{z}_{ij}\frac{e^{\tilde{\lambda}_j\rho}}{\rho}\left(\int_0^\rho te^{-\tilde{\lambda}_j t}h_{jk}B_k^{\lessgtr}(t)\mathrm{d}t - \beta_j^{\lessgtr}\right).$$

Here the constants $\beta_j^{\lessgtr}$ are determined by the boundary and matching conditions and the inhomogeneous functions are denoted by B_k, with k the equation number. To determine the value of the weighting constants h_{jk} one needs to invert the $2\mathrm{n}^2$ constraint equations

$$\sum_{j=1}^{2n} \tilde{z}_{ij}h_{jk}\tilde{\lambda}_j = \delta_{ik} \tag{4.87}$$

$$\sum_{j=1}^{2n} \tilde{z}_{ij}h_{ik} = 0. \tag{4.88}$$

The above equations are all matrix equations of rank n which result in $2\mathrm{n}^2$ constraints in total. This of course is the solution for the case where the mass matrix is approximated by a series of step functions. If we wish to handle its full space–time varying profile, this can be achieved by treating corrections to the above solution also as a perturbation. Let us relabel the above solution ϵ_i^0 and substitute $\epsilon_i = \epsilon_i^0 + \delta\epsilon_i + \cdots$ into the differential equations denoting the deviation from the step function approximation of the mass matrix as $\eta_{ij}(\rho)$. If we keep only first-order terms we once again find we have a linear set of differential equations

$$\begin{aligned} \frac{\partial^2\epsilon_i}{\partial\rho^2} + \frac{2}{\rho}\frac{\partial\epsilon_i}{\partial\rho} - \left.\frac{\partial^2 V(\phi)}{\partial\phi_i\partial\phi_j}\right|_A \epsilon_j = B_i(\rho) \\ \frac{\partial^2\delta\epsilon_i + \epsilon_i^0}{\partial\rho^2} + \frac{2}{\rho}\frac{\partial\delta\epsilon_i + \epsilon_i^0}{\partial\rho} - \left(\delta\epsilon_j + \epsilon_j^0\right)\left(\bar{M}_{ij} + \eta_{ij}\right) = B_i(\rho). \end{aligned} \tag{4.89}$$

We can observe that ϵ_i^0 by definition is the solution to the initial differential equations with mass matrix $\bar{M}_{ij}$. This allows a dramatic simplification in the above equation

$$\begin{aligned} \frac{\partial^2\delta\epsilon_i}{\partial\rho^2} + \frac{2}{\rho}\frac{\partial\delta\epsilon_i}{\partial\rho} - \delta\epsilon_j\bar{M}_{ij} - \eta_{ij}\epsilon_j^0 = 0 \\ \frac{\partial^2\delta\epsilon_i}{\partial\rho^2} + \frac{2}{\rho}\frac{\partial\delta\epsilon_i}{\partial\rho} - \delta\epsilon_j\bar{M}_{ij} = \eta_{ij}\epsilon_j^0. \end{aligned} \tag{4.90}$$

Once again we have a network of coupled linear differential equations which has the same form, and therefore the same form of the solution, as the differential equations for ϵ_i^0. Therefore to write the solution to the above equation we merely replace B_i with $\eta_{ij}\epsilon_j^0$. To obtain the bounce solution, simply calculate the series ϵ_i with corrections $\delta\epsilon_i$ up to a desired tolerance.

4.7.1 Observations on convergence

The semi-analytic approach has similarities to a functional version of Newton's method. This raises the question of whether it might face analogous convergence issues. That is:

1. Will convergence be slow if the guess is far from the true solution?
2. Are there cases where the perturbations oscillate rather than converge? That is can $\epsilon^{(n)}(\rho) \approx -\epsilon^{(n+1)}(\rho)$?
3. Is there an analogous problem to where the perturbation diverges like it does in Newton's method when the derivative is zero?
4. Can you converge toward the wrong solution due to being in the wrong basin of attraction?

The semi-analytic perturbative method is immune from any of these problems, except the fourth, which is a general problem for all known algorithms which attempt to find bubble wall profiles. We previously made an argument that this method is generically free from the first problem as the choice of the initial ansatz generically ensures perturbations bounded by 1. The remaining two issues are more difficult to deal with.

We begin by discussing the possibility of oscillating solutions. Let the ansatz with a n perturbative corrections be denoted by

$$A_i^{(n)} = A_i^{(0)} + \sum_{k=1}^{n-1} \epsilon_i^{(k)}.$$

In the above we denote the initial ansatz as $A_i^{(0)}$ with corrections $\epsilon_i^{(k)}$. Let us assume that for a particular field ϕ_i, each correction oscillates at correction n, such that $\epsilon_i^{(n)} = -\epsilon_i^{(n+1)}$. We can then write the equations of motion for $A_i^{(n+2)} = A_i^{(n)} + \epsilon_i^{(n)} + \epsilon_i^{(n+1)}$ before Taylor expanding as

$$\begin{aligned} &\frac{\partial^2\left[A_i^{(n)} + \epsilon_i^{(n)} + \epsilon_i^{(n+1)}\right]}{\partial\rho^2} + \frac{2}{\rho}\frac{\partial\left[A_i^{(n)} + \epsilon_i^{(n)} + \epsilon_i^{(n+1)}\right]}{\partial\rho} + \left.\frac{\partial V(\phi)}{\partial\phi_i}\right|_{\left\{A_k^{(n)}+\epsilon_k^{(n)}+\epsilon_k^{(n+1)}\right\}} \\ &= \frac{\partial^2 A_i^{(n)}}{\partial\rho^2} + \frac{2}{\rho}\frac{\partial A_i^{(n)}}{\partial\rho} + \left.\frac{\partial V}{\partial\phi_i}\right|_{\left\{A_1^{(n+2)}, \ldots, A_i^{(n)}, A_{i_1}^{(n+2)}, \ldots\right\}} = 0. \end{aligned} \tag{4.91}$$

So we see that in the case of the single field an oscillating solution implies an exact solution to the equations of motion. Let us now turn to the multi-field case. The other fields $\phi_{j\neq i}$ will converge without oscillating corrections. Therefore the corrections to change to the derivative of the tree-level potential will decay and thus will converge similarly to the single-field case. In the unlikely event that multiple fields start to have oscillating corrections, convergence could become slow. However, since the equations are still coupled one can make a similar argument to the above to see that convergence is still likely.

The other issue that needs to be checked is what happens when the mass matrix becomes zero. In this case the solution to the differential equations for the perturbations are

$$\frac{\partial^2 \epsilon_i}{\partial \rho^2} + \frac{2}{\rho}\frac{\partial \epsilon_i}{\partial \rho} = B_i(\rho).$$

This is easily solved as

$$\epsilon_i = \beta_0 + \frac{\beta_{-1}}{\rho} + \int_0^{\rho} \frac{\mathrm{d}y}{y^2} \int_0^{y} x^2 B(x) \mathrm{d}x,$$

where β_1 and β_{-0} will both be zero when the mass matrix is zero for all space–time points. Alternatively, if the mass matrix is singular one needs to take the singular value composition (or alternatively one can use a particular arrangement of the step functions used to approximate the mass matrix such that it is not singular in any region). All these considerations are of course avoided if one solves the differential equations governing the perturbations numerically.

4.7.2 A numerical example

The semi-analytic approach has a lot of steps, some of which are analytically intense. It is useful to demonstrate how it works using an example. Let us consider the two-field potential which was described in [32],

$$V(x, y) = \left(x^2 + y^2\right)\left[1.8(x-1)^2 + 0.2(y-1)^2 - \delta\right].$$

In the above, when the parameter $\delta = 0.4$ the path deviates a long way from the straight line ansatz so this is a sturdy test of the perturbative approach. The first step is to create the ansatz which is a straight line trajectory in field space. We therefore make the rotation in field space $(x, y) \to (u, v)$, where the global minimum is in the u direction. To reduce to a single-field problem we then simply set v to zero in the effective potential. We next want the global minimum to be at $u = 1$. This is achieved by a rescaling (and in general a shift) followed by division by $|E|$ which gives

$$\frac{V(u, 0)}{|E|} = 0.36u^2 - u^3 + 0.57u^4.$$

Our analytic formulas can now be directly used to make the ansatz. Doing so and then reversing the rescaling and rotation, the ansatz in the (x, y) basis is

$$x(\rho) = 1.046\left(1 - \tanh\left[\frac{\rho - 0.437}{1}\right]\right) \tag{4.92}$$

$$y(\rho) = 1.663\left(1 - \tanh\left[\frac{\rho - 0.437}{1}\right]\right). \tag{4.93}$$

Here we have the wall width being unity as a coincidence due to the rescaling. A useful trick to help convergence is to sanitize the above ansatz such that the derivative is zero when $\rho \to 0$. This is achieved by making a small modification to the ansatz:

$$\delta x(\rho) = \left.\frac{\partial x}{\partial \rho}\right|_{\rho=0} \exp\left[-\left.\frac{\partial x}{\partial \rho}\right|_{\rho=0} \rho\right] \tag{4.94}$$

$$\delta y(\rho) = \left.\frac{\partial y}{\partial \rho}\right|_{\rho=0} \exp\left[-\left.\frac{\partial y}{\partial \rho}\right|_{\rho=0} \rho\right]. \tag{4.95}$$

Each trajectory of $(x(\rho), y(\rho))$ is calculated for each successive perturbation in figure 4.11 and compared to the true solution. It takes three perturbations to converge to the real solution. In figure 4.12 the space–time dependence of the x and y fields is shown with each successive perturbation beginning with the single field ansatz. Finally, figure 4.13 shows the convergence of the perturbation $B_x(\rho)$, plotted for successive perturbations of $x(\rho)$ starting with the ansatz. Incredibly, the magnitude of the error reduces by almost an order of magnitude with each perturbation.

4.7.3 Wall width and variation in β

In later chapters we will find that if nature has more than one Higgs field, it is possible to have a resonance in baryon number production. The peak of the

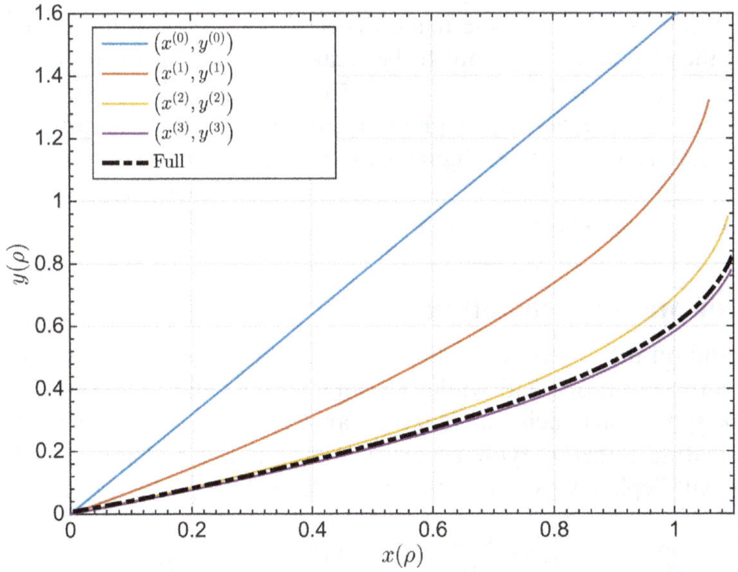

Figure 4.11. The bounce solution presented in field space for successive perturbations. The initial ansatz is the solid straight line while each perturbation is given by the solid curved lines. The numerical result from [32] is given by the dashed curve. Reproduced from [34].

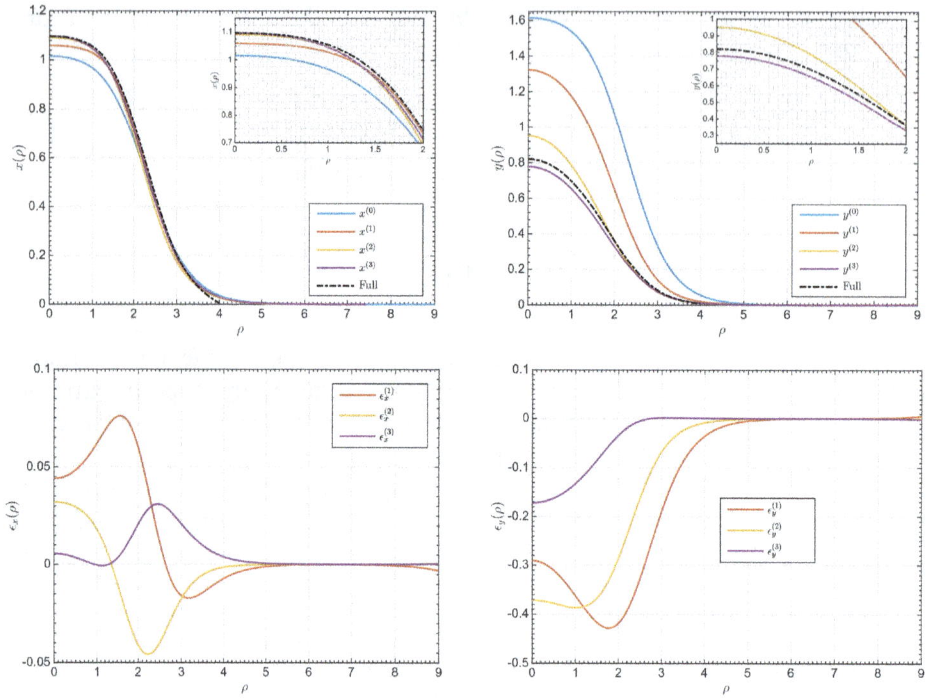

Figure 4.12. The single field ansatz and the first three perturbations for x and y. The initial ansatzes for x and y are in the top left and right panels, respectively. The bottom left and right panels show the first three perturbations for x and y, respectively. Reproduced from [34].

resonance occurs when a particle has a degenerate mass with a virtual state. The strength of the resonance turns out to be weakly dependent on the wall width and linearly proportional to $\Delta\beta$, where $\tan\beta(T) = v_u(T)/v_d(T)$ and $\Delta\beta$ describes the maximum variation of β during the phase transition [37, 38]. Generally it needs to be calculated numerically, however, figure 4.14 shows a schematic of the tunneling path of the phase transition through two directions in field space. Generically if one has more directions in field space it becomes easier for $\Delta\beta$ to be large.

4.8 Baryon washout condition

Let us expand on the criteria of a strongly first-order phase transition now that we have the sufficient tools. CP violating interactions with the bubble wall bias electroweak sphalerons such that baryons are created more frequently than anti-baryons, creating a net baryon asymmetry within the new phase. This baryon asymmetry will deplete within the new phase according to the rate equation

$$\frac{\mathrm{d}n_B}{\mathrm{d}t} = -k(T)n_B^n \tag{4.96}$$

for a temperature-dependent rate constant k, and n is the order of the phase transition. For a first-order phase transition n_B decays exponentially according to

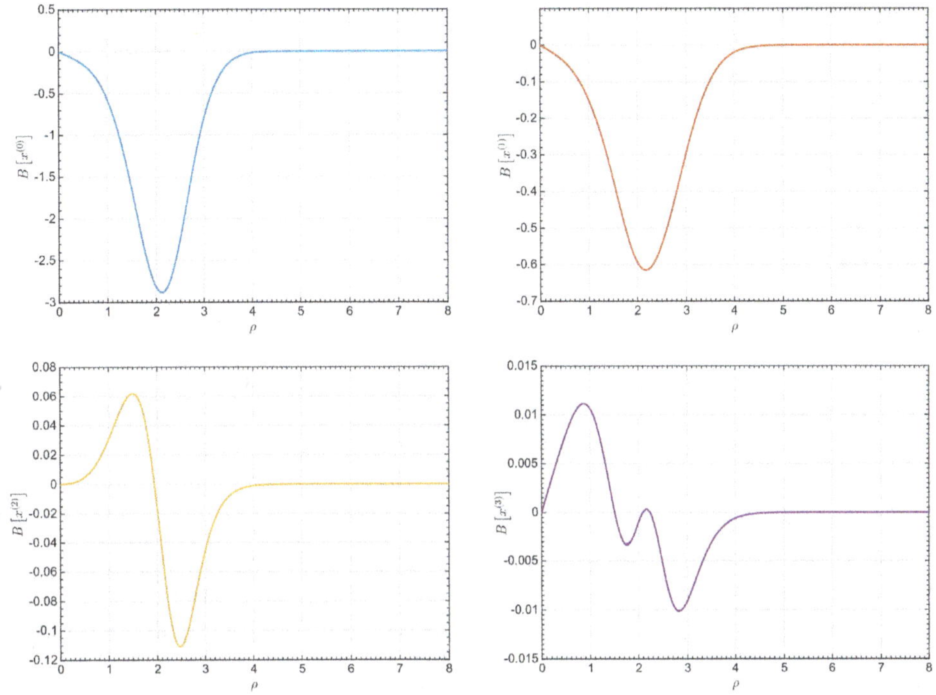

Figure 4.13. The error function B for the single field ansatz for x and the first three perturbations. Reproduced from [34]

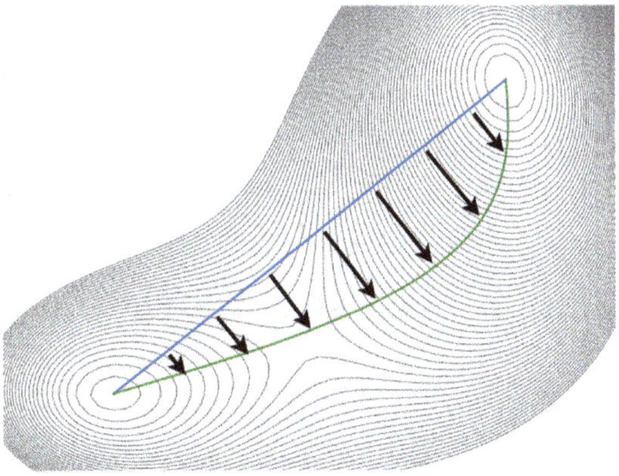

Figure 4.14. An example of how $\tan\beta$ can vary during a phase transition. The figure is a contour plot in field space of the effective potential of a two scalar field system. The tunneling path from one minimum to another is a curved path in field space so $\tan\beta$ has a maximum and minimum value. The difference in the maximum and minimum value of β during the phase transition gives $\Delta\beta$. Reproduced from [32].

some half life $\ln 2/k(T)$ whereas a second-order phase transition decays inversely with time with a time-dependent half life. Naively at this point it might appear that a second-order phase transition is superior to a first-order phase transition when it comes to baryon number preservation. However, it is the difference in the half life between the two types of phase transitions that makes a strongly first-order phase transition necessary. The rate constant is just the sphaleron rate at high temperature which is given by some prefactor times an exponential of the strength of the phase transition

$$k(T) = -\frac{13n_f}{2VT^3}\frac{\omega_-}{2\pi}(\mathcal{NV})_{\text{rot}}(\mathcal{NV})_{\text{tr}}\left(\frac{v_C}{gT_C}\right)^3 \mathrm{e}^{\frac{v_C(T_C)}{T_C}\frac{4\pi B\left(\frac{\lambda}{g}\right)}{g}}\kappa. \tag{4.97}$$

We can define a washout factor, S, as the ratio of the baryon asymmetry at the end of the phase transition, $n_B(t_f)$ compared to its value at the start, $n_B(t_i)$. The process of calculating the initial baryon asymmetry we will leave to the later chapters but for now we can make some qualitative observations on the behavior of S as we modify v_C/T_C. A simple integration of the rate equation reveals a double exponential dependence on the strength of the phase transition

$$S \sim \exp \mathrm{e}^{-\frac{v_C}{T_C}}. \tag{4.98}$$

So $S \to 1$ very quickly as we increase v_C/T_C. Suppose we have a model that can produce a baryon asymmetry e^X times the baryon asymmetry of the Universe at the time of the phase transition for a certain set of parameters. We then have a bound on the washout condition of

$$\frac{n_B(t_f)}{n_B(0)} > \mathrm{e}^{-X}. \tag{4.99}$$

Assuming that $k(T)$ is weakly dependent on time one can derive the bound

$$\frac{4\pi B}{g}\frac{v_C}{T_C} - 6\ln\frac{v_C}{T_C} > -\ln X - \ln\frac{t_f}{t_i} + \ln\frac{13n_f}{2}\frac{\omega_-}{2\pi}(\mathcal{NV})_{\text{rot}}(\mathcal{NV})_{\text{tr}} + \ln\kappa. \tag{4.100}$$

For $X \approx 1$ and $T_C \approx 100$ one find that the above reduces to

$$\frac{v_C}{T_C} \gtrsim 1 \tag{4.101}$$

with theoretical uncertainties of about ~0.25 [17].

References

[1] Weinberg S 1974 Gauge and global symmetries at high temperature *Phys. Rev.* D **9** 3357

[2] Arnold P and McLerran L 1987 Sphalerons, small fluctuations, and baryon-number violation in electroweak theory *Phys. Rev.* D **36** 2

[3] Shaposhnikov M E 1988 Structure of the high temperature gauge ground state and electroweak production of the baryon asymmetry *Nucl. Phys.* B **299** 4

[4] Arnold P and McLerran L 1988 The sphaleron strikes back: A response to objections to the sphaleron approximation *Phys. Rev.* D **37** 4

[5] Quiros M 1999 Finite temperature field theory and phase transitions arxiv: 9901312v1 (hep-ph)

[6] Das A 1997 *Finite Temperature Field Theory* (Singapore: World Scientific)

[7] Schwinger J 1961 Brownian motion of a quantum oscillator *J. Math. Phys.* **2** 407

[8] Mahanthappa K T 1962 Multiple production of photons in quantum electrodynamics *Phys. Rev.* **126** 329

[9] Bakshi P M and Mahanthappa K T 1963 Expectation value formalism in quantum field theory I *J. Math. Phys.* **4** 1

[10] Bakshi P M and Mahanthappa K T 1963 Expectation value formalism in quantum field theory II *J. Math. Phys.* **4** 12

[11] Keldysh L V 1965 Diagram technique for nonequilibrium processes *Sov. Phys.–JETP* **20** 1018

[12] Craig R A 1968 Perturbation expansion for real-time Green's functions *J. Math. Phys.* **9** 605

[13] Chou K C, Su Z B, Hao B L and Yu L 1985 Equilibrium and nonequilibrium formalisms made unified *Phys. Rep.* **118** 1

[14] Millington P and Apostolos P 2013 Perturbative nonequilibrium thermal field theory *Phys. Rev.* D **88** 085009

[15] Peskin M E and Shroeder D V 1995 *An Introduction to Quantum Field Theory* (Boulder, CO: Westview)

[16] Dolan L and Jackiw R 1974 Symmetry behavior at finite temperature *Phys. Rev.* D **9** 3320

[17] Patel H H and Ramsey-Musolf M J 2011 Baryon washout, electroweak phase transition, and perturbation theory *J. High Energy Phys.* JHEP07(2011)029

[18] Nielsen N K 1975 On the gauge dependence of spontaneous symmetry breaking in gauge theories *Nucl. Phys.* B **101** 1

[19] Fukuda R and Kugo T 1976 Gauge invariance in the effective action and potential *Phys. Rev.* D **13** 12

[20] Fendley P 1987 The effective potential and the coupling constant at high temperature *Phys. Lett.* B **196** 2

[21] Espinosa J R, Quirós M and Zwirne F 1992 On the phase transition in the scalar theory *Phys. Lett.* B **291** 115–24

[22] Carson L and McLerran L D 1990 Approximate computation of the small-fluctuation determinant around a sphaleron *Phys. Rev.* D **41** 647

[23] Carson L, Li X, McLerran L D and Wang R-T 1990 Exact computation of the small-fluctuation determinant around a sphaleron *Phys. Rev.* D **42** 2127

[24] Higgs P W 1964 Broken symmetries and the masses of gauge bosons *Phys. Rev. Lett.* **13** 16

[25] Coleman S 1977 Fate of the false vacuum: semiclassical theory *Phys. Rev.* D **15** 10

[26] Callan C G Jr and Coleman S 1977 Fate of the false vacuum. II. First quantum corrections *Phys. Rev.* D **16** 6

[27] Coleman S and De Luccia F 1980 Gravitational effects on and of vacuum decay *Phys. Rev.* D **21** 12

[28] McLerran L, Shaposhnikov M, Turok N and Voloshin M 1991 Why the baryon asymmetry of the Universe is $\approx 10^{-10}$ *Phys. Lett.* B **256** 477–83

[29] Dine M, Huet P and Singleton R Jr 1992 Baryogenesis at the electroweak scale *Nucl. Phys.* B **375** 625–48

[30] Moreno J M, Quiros M and Seco M 1998 Bubbles in the supersymmetric standard model *Nucl. Phys.* B **526** 489–500

[31] Profumo S, Ubaldi L and Wainwright C 2010 Singlet scalar dark matter: monochromatic gamma rays and metastable vacua *Phys. Rev.* D **82** 123514

[32] Wainwright C L 2012 CosmoTransitions: computing cosmological phase transition temperatures and bubble profiles with multiple fields *Comput. Phys. Commun.* **183** 2006–13

[33] Dine M, Leigh R G, Huet P Y, Linde A D and Linde D A 1992 Towards the theory of the electroweak phase transition *Phys. Rev.* D **46** 550

[34] Akula S, Balázs C and White G A 2016 Semi-analytic techniques for calculating bubble wall profiles arXiv: 1608.00008(hep-th)

[35] John P 1999 Bubble wall profiles with more than one scalar field: a numerical approach *Phys. Lett.* B **452** 221–6

[36] White G A 2016 General analytic methods for solving coupled transport equations: from cosmology to beyond *Phys. Rev.* D **93** 4

[37] Carena M, Quirós M, Riott A, Vilja I and Wagner C E M 1997 Electroweak baryogenesis and low energy supersymmetry *Nucl. Phys.* B **503** 1

[38] Moreno J M, Quiros M and Seco M 1998 Bubbles in the supersymmetric Standard Model *Nucl. Phys.* B **526** 1

Chapter 5

CP violation

The third Sakharov condition that needs to be satisfied is violation of C and CP. The Standard Model has CP violation through the CKM matrix—a CP violating phase in Higgs–quark Yukawa interactions cannot be completely rotated away through field redefinitions as one CP violating phase must remain [1]. The EDMs (electric dipole moments) produced by the CKM matrix are well below experimental lower bounds [2] and produce far too feeble a CP violation to produce the observed baryon asymmetry. This means there is plenty of experimental room for new sources of CP violation necessary for baryogenesis to hide in. The way of testing CP violation experimentally is through the search for permanent electric dipole moments in electrons, protons, and neutrons [3, 4]. This means that tests involve matter rather than anti-matter, so CPT invariance is assumed. Also, in general, the calculations involve zero temperature two-loop (or higher) calculations that involve the calculation of Wilson coefficients. The details of this can be found in many sources outside this book, so we only give details of the set up of the problem and the big picture.

The summary of the logic is as follows. First systematically list all diagrams that lead to CPV operators of the form given in tables 5.1 and 5.2 once heavy BSM (beyond the standard model) fields are integrated out. Example of such diagrams for the MSSM are given in figure 5.1. This gives an effective operator for the electron EDM (electric dipole moment). For neutron and proton EDMs one needs to run from the weak scale to below the scale Λ_χ and calculate the appropriate Wilson coefficients that arise from systematically calculating appropriate matrix elements from the relevant terms in the effective Lagrangian.

The non-relativistic expression for an electric dipole moment for a spin-two particle is

$$H = -d\frac{\vec{E} \cdot \vec{s}}{|s|}, \tag{5.1}$$

doi:10.1088/978-1-6817-4457-5ch5

Table 5.1. List of dimension six CPV operators excluding four fermion interactions. List reproduced from [6].

Pure gauge		Gauge Higgs		Gauge Higgs fermion	
$\mathcal{O}_{\tilde{G}}$	$f^{abc} G^a_{\mu\nu} \tilde{G}^{b,\nu\beta} G^{c\mu}_{\beta}$	$\mathcal{O}_{\varphi\tilde{G}}$	$\varphi^\dagger\varphi \tilde{G}^a_{\mu\nu} G^{a,\,\mu\nu}$	$\mathcal{O}_{uG}$	$(\overline{Q}\sigma^{\mu\nu}\lambda^a u_R)\tilde{\varphi} G^a_{\mu\nu}$
$\mathcal{O}_{\tilde{W}}$	$\epsilon^{ijk} W^i_{\mu\nu} \tilde{W}^{j,\nu\beta} W^{k\mu}_{\beta}$	$\mathcal{O}_{\varphi\tilde{W}}$	$\varphi^\dagger\varphi \tilde{W}^i_{\mu\nu} W^{i,\mu\nu}$	$\mathcal{O}_{fW}$	$(\overline{f}\,\sigma^{\mu\nu}\tau^i f_R)\Phi W^i_{\mu\nu}$
		$\mathcal{O}_{\varphi\tilde{B}}$	$\varphi^\dagger\varphi \tilde{B}_{\mu\nu} B^{\mu\nu}$	$\mathcal{O}_{dG}$	$(\overline{Q}\sigma^{\mu\nu}\lambda^a d_R)\tilde{\varphi} G^a_{\mu\nu}$
		$\mathcal{O}_{\varphi\tilde{W}B}$	$\varphi^\dagger\varphi\tau^i \tilde{W}^i_{\mu\nu} B^{\mu\nu}$	$\mathcal{O}_{fB}$	$(\overline{f}\,\sigma^{\mu\nu} f_R)\Phi B_{\mu\nu}$

Table 5.2. List of dimension-six CPV four-fermion operators. List reproduced from [6].

Four fermion interactions	
$\mathcal{O}_{ledq}$	$\overline{L}^j e_R \overline{d}_R Q^j$
$\mathcal{O}_{quqd}$	$\overline{Q}^i u_R \epsilon_{ij} \overline{Q}^j d_R$
$\mathcal{O}^8_{quqd}$	$\overline{Q}^i \lambda^a u_R \epsilon_{ij} \overline{Q}^j \lambda^a d_R$
$\mathcal{O}_{lequ}$	$\overline{L} e_R \epsilon_{ij} \overline{Q} u_R$
$\mathcal{O}^3_{lequ}$	$\overline{L}\sigma_{\mu\nu} e_R \epsilon_{ij} \overline{Q}\sigma^{\mu\nu} u_R$

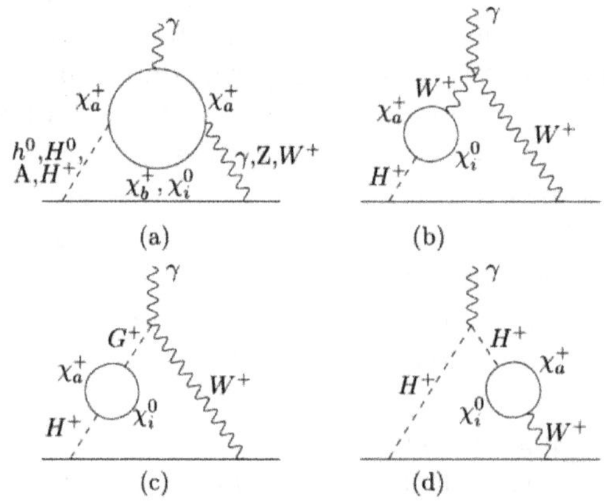

Figure 5.1. Some example operators that contribute to EDMs involving neutralino and charginos in the MSSM. Reproduced from [5].

which has the relativistic generalization [7]

$$\mathcal{L}_{\text{EDM}} = -d\frac{\mathrm{i}}{2}\overline{\psi}\sigma^{\mu\nu}\gamma_5\psi F_{\mu\nu}. \tag{5.2}$$

One typically uses the framework of effective field theory to calculate contributions to EDMs [8–9]. To calculate the observable EDM we need to consider the appropriate energy scale for a relevant experiment be it the QCD, nuclear, or atomic scale. One can then write out the set of higher dimensional operators that contribute to EDM, organizing them into powers of $1/\Lambda$ with the leading order expected to be dominant. At the nuclear scale the observables relevant to CP violation are the electron EDM d_e, the proton EDM d_p and the neutron EDM d_n, as well as electron–nucleon and nucleon–nucleon interactions with Lagrangians. These latter interactions are the parity violating and time reversal violating (PVTV) πNN interaction, the nuclear Schiff moment, and the PVTV electron–nucleus interaction.

For an energy scale above 1 GeV the Lagrangian organized into powers of $1/\Lambda$ is

$$\mathcal{L} = \mathcal{L}_{D=4} + \frac{1}{\Lambda}\mathcal{L}_{D=5} + \frac{1}{\Lambda^2}\mathcal{L}_{D=6} + O\left(\frac{1}{\Lambda^3}\right). \tag{5.3}$$

There is only one dimension-four operator which is topologically induced

$$\mathcal{L}_{D=4} = \frac{\alpha_s}{8\pi}\bar{\theta}G_{\mu\nu}\tilde{G}^{\mu\nu}. \tag{5.4}$$

This term produces far too much CP violation unless $\bar{\theta}$ is vanishingly small. This is known as the strong CP problem, with the most popular solution being the axion. We will assume that $\bar{\theta} = 0$ and ignore the dimension-four contribution. Naively the dimension-five terms are just $\mathcal{L}_{\text{EDM}}$, however for Standard Model particle content, gauge invariance forbids such a term without an additional Higgs field. The other dimension-six operators are listed in tables 5.1 and 5.2.

The gauge-Higgs–fermion interactions lead to terms of the form $\mathcal{L}_{\text{EDM}}$. After replacing the Higgs with its vacuum expectation value we obtain

$$\mathcal{O}_{qG} \to -\mathrm{i}\sum_q \frac{g_3 d_{\tilde{q}}}{2}\bar{q}\sigma^{\mu\nu}\gamma_5\lambda^a q G^a_{\mu\nu} \tag{5.5}$$

$$\mathcal{O}_{fW} \to -\mathrm{i}\sum_f \frac{d_f}{2}\bar{f}\,\sigma^{\mu\nu}\gamma_5 f F_{\mu\nu} \tag{5.6}$$

with electric dipole moments

$$d_{\tilde{q}} = -\sqrt{2}\frac{v}{\Lambda^2}\,\mathrm{Im}\left[C_{qG}\right] \tag{5.7}$$

$$d_f = -\sqrt{2}\,e\frac{v}{\Lambda^2}\,\mathrm{Im}\left[C_{f\gamma}\right] \tag{5.8}$$

where C_x are the Wilson coefficients.

Next one needs to relate the $D = 6$ CPV interactions to observables at the hadronic scale. The most useful framework one uses to calculate observables at this scale is heavy baryon chiral perturbation theory. Denoting the BSM scale as Λ and the chiral symmetry breaking scale as Λ_χ, one then produces an expansion in powers of Λ_χ/Λ. The four-fermion operators lead to effective interactions between electrons and nucleons given by

$$\mathcal{L}_{eN} = \bar{e}i\gamma_5 e\bar{N}\left(C_S^0 + iC_S^1\tau_3\right)N + \bar{e}e\bar{N}i\gamma_5\left(C_P^0 + iC_P^1\tau_3\right)N \\ + \epsilon_{\mu\nu\alpha\beta}\bar{e}\sigma^{\mu\nu}e\bar{N}\sigma^{\alpha\beta}\left(C_T^0 + iC_T^1\tau_3\right)N, \tag{5.9}$$

whereas the PVTV interactions involving pions and nucleons responsible for the nuclear Schiff moment are

$$\mathcal{L}_{\pi NN} = g_{\pi NN}^0\bar{N}\tau^a N\pi^a + g\pi NN^1\bar{N}N\pi^0 + g_{\pi NN}^2\left(\bar{N}\tau^a N\pi^a - 3\bar{N}\tau^3 N\pi^0\right). \tag{5.10}$$

Finally the Lagrangian for the nucleon EDM at the hadronic scale defined in the nucleon rest frame is given by

$$\mathcal{L}_N = -2\bar{N}\left(d_0 + d_1\tau^3\right)\sigma^i NF^{0i}. \tag{5.11}$$

The hadronic couplings $d_p = d_0 + d_1$, $d_n = d_0 - d_1$, C_S, C_T, C_P, and $g_{\pi NN}^i$ are computed by calculating the matrix elements of the operators given in tables 5.1 and 5.2. If we ignore $\bar{\theta}$ which arises from the four dimensional CPV operator, the hadronic couplings have the form

$$\begin{aligned} d_{n,p} &= \frac{v^2}{\Lambda^2}\sum\beta_{n,p}\ \mathrm{Im}\left[C_{n,p}\right] \\ g_{\pi NN}^i &= \frac{v^2}{\Lambda^2}\sum\gamma\ \mathrm{Im}\left[C_{piNN}\right] \\ C_x &= \frac{v^2}{\Lambda^2}\sum\delta\ \mathrm{Im}[C_C]. \end{aligned} \tag{5.12}$$

Experimental probes of EDMs look at paramagnetic atoms and polar molecules, which are most sensitive to d_e and C_S [7, 10, 11], diamagnetic atoms which are most sensitive to C_T and $g_{\pi NN}^{0,1}$, and neutron EDMs which are most sensitive to d_n and $g_{\pi NN}$ [12]. To conclude we give the current experimental bounds on the neutron and electron EDMs compared to the Standard Model value [2–4]

$$\begin{aligned} d_n^{\mathrm{SM}} &= 10^{-31}ecm \qquad & d_n^{\exp} &< 3.0 \times 10^{26}ecm \\ d_e^{\mathrm{SM}} &= 10^{38}ecm \qquad & d_e^{\exp} &< 8.7 \times 10^{-29}ecm, \end{aligned} \tag{5.13}$$

whereas the bounds on c_X and $g_{\pi NN}$ are

$$\begin{aligned} C_S &< 4.5 \times 10^{-7} \\ C_T &< 2 \times 10^{-6} \\ g_{\pi NN}^0 &< 8 \times 10^{-9} \\ g_{\pi NN}^1 &< 1 \times 10^{-9}. \end{aligned} \tag{5.14}$$

References

[1] Peskin M E and Shroeder D V 1995 *An Introduction to Quantum Field Theory* (Boulder, CO: Westview)

[2] Dar S 2000 The neutron EDM in the SM: a review arXiv: hep-th-0008248

[3] Pendlebury J M *et al* 2015 Revised experimental upper limit on the electric dipole moment of the neutron *Phys. Rev.* D **92** 092003
[4] Baron J *et al* The ACME Collaboration 2014 Order of magnitude smaller limit on the electric dipole moment of the electron *Science* **343** 269–72
[5] Li Y, Profumo S and Ramsey-Musolf M 2008 Higgs-higgsino-gaugino induced two loop electric dipole moments *Phys. Rev.* D **78** 7
[6] Engel J, Ramsey-Musolf M J and Van Kolck U 2013 Electric dipole moments of nucleons, nuclei, and atoms: the Standard Model and beyond *Prog. Part. Nucl. Phys.* **71** 21–74
[7] Pospelov M and Ritz A 2005 Electric dipole moments as probes of new physics *Ann. Phys.* **318** 1
[8] Weinberg S 1976 Gauge theory of CP nonconservation *Phys. Rev. Lett.* **37** 11
[9] Barr S M and Zee A 1990 Electric dipole moment of the electron and of the neutron *Phys. Rev. Lett.* **65** 1
[10] Dzuba V A, Flambaum V V and Harabati C 2011 Relations between matrix elements of different weak interactions and interpretation of the parity-nonconserving and electron electric-dipole-moment measurements in atoms and molecules *Phys. Rev.* A **84** 5
[11] Jung M 2013 A robust limit for the electric dipole moment of the electron *J. High Energy Phys* 168JHEP05(2013)
[12] Chupp T and Ramsey-Musolf M 2015 Electric dipole moments: a global analysis *Phys. Rev.* C **91** 3

Chapter 6

Particle dynamics during a phase transition

Electroweak baryogenesis occurs during the electroweak phase transition, a process that happens when the Universe is at a temperature of $\mathcal{O}(100)$ GeV in most models. As we will see later, this is probably the latest time that the baryon asymmetry can be produced since other mechanisms tend to produce the BAU before the electroweak phase transition.[1] Earlier we introduced the CTP formalism for handling the effective potential at finite temperature. However, we also need a way of describing the particle dynamics in a far-from-equilibrium setting [2]. In particular, conventionally equilibrium finite temperature QFT misses the substantial non-Markovian nature of such particle dynamics. In particular, these memory effects can cause a substantial boost to the baryon asymmetry through resonant effects [3–4]. In this section our propagators become functions of space–time-dependent chemical potentials due to the modification of the Fermi and Bose distributions [5–6]

$$n_{\mathrm{F,B}}(p_0) \to \frac{1}{\exp\left[\beta(p_0 + \mu)\right] \pm 1}. \tag{6.1}$$

We will show how the space–time-dependent chemical potentials govern the evolution of particle number currents. Eventually the chemical potentials themselves will be related to the number densities.

6.1 Particle current divergences and self-energy

Within the CTP formalism, divergences or particle currents can be related to the particles' self-energies [7]. We work with bosons to start with. Recall that our choice

[1] A notable exception to this is post-sphaleron baryogenesis [1].

doi:10.1088/978-1-6817-4457-5ch6

of time contour results in four independent Green's functions rather than the usual two. These can be written in matrix form for convenience:

$$G(x, z) \equiv \begin{pmatrix} G^t(x, z) & -G^<(x, z) \\ G^>(x, z) & -G^{\bar{t}}(x, z) \end{pmatrix}. \tag{6.2}$$

The Schwinger–Dyson equation is then a matrix of equations.

$$\begin{aligned} \tilde{G}(x, y) &= \tilde{G}^0(x, y) + \int \mathrm{d}^4w \int \mathrm{d}^4z \tilde{G}^0(x, w)\tilde{\Sigma}(w, z)\tilde{G}(z, y) \\ \tilde{G}(x, y) &= \tilde{G}^0(x, z) + \int \mathrm{d}^4w \int \mathrm{d}^4z \tilde{G}(x, w)\tilde{\Sigma}(w, z)\tilde{G}^0(z, y). \end{aligned} \tag{6.3}$$

Here $\tilde{\Sigma}(x, z)$ is a matrix of self-energies corresponding to each propagator. Let us act with the Klein–Gordon operator, $K(x) \equiv \Box_x + m^2$, on both Schwinger–Dyson equations, remembering that the Klein–Gordon operator acting on $\tilde{G}^0$ gives a delta function. We will also assume, as a simplification and therefore an approximation, that the mass associated with the Klein–Gordon operator is not space–time-dependent during the phase transition. The result can be expressed in matrix components denoted by the subscript *ab*,

$$\begin{aligned} K(x)\tilde{G}(x, y)_{ab} &= -\mathrm{i}\delta^4(x - y) - \mathrm{i}\int \mathrm{d}^4z\tilde{\Sigma}(x, z)_{ac}\tilde{G}(z, y)_{cb} \\ K(y)\tilde{G}(x, t)_{ab} &= -\mathrm{i}\delta^4(x - y) - \mathrm{i}\int \mathrm{d}^4z\tilde{G}(x, y)_{ac}\tilde{\Sigma}(z, y)_{cb}. \end{aligned} \tag{6.4}$$

Let us subtract the two equations and take the limit

$$\lim_{x\to y}\left[K(x)\tilde{G}(x, y)_{ab} - K(y)\tilde{G}(x, y)_{ab}\right]. \tag{6.5}$$

Consider the (1, 1) term of the above limit. The delta functions cancel and the resulting integrand, *I*, is

$$\begin{aligned} I = \lim_{x\to y}&\left[\theta(x_0 - z_0)\Sigma^>(x, z) + \theta(z_0 - x_0)\Sigma(x, z)\right] \\ &\times\left[\theta(z_0 - y_0)G^>(z, y) + \theta(y_0 - z_0)G^<(z, y)\right] \\ &-\left[\theta(x_0 - z_0)G^>(x, z) + \theta(z_0 - x_0)G^<(x, z)\right] \\ &\times\left[\theta(z_0 - y_0)\Sigma^>(z, y) + \theta(y_0 - z_0)\Sigma^<(z, y)\right] \\ &-\Sigma^>(x, z)G^<(z, y) + G^>(x, z)\Sigma^<(z, y). \end{aligned} \tag{6.6}$$

Using the identities that $\theta(x_0) + \theta(-x_0) = 1$ and $\theta(x_0)^2 = \theta(x_0)$ we can simplify the integrand to first remove products of step functions and then write everything proportional to a single step function,

$$
\begin{aligned}
I = &[\theta(x_0 - z_0) - 1]\Sigma^{>}(x, z)G^{<}(z, x) + \theta(z_0 - x_0)\Sigma^{<}(x, z)G^{<}(z, x) \\
&\times [1 - \theta(x_0 - z_0)]G^{>}(x, z)\Sigma^{<}(z, x) - \theta(z_0 - x_0)G^{<}(x, z)\Sigma^{>}(z, x) \\
= &- \theta(z_0 - x_0)\Sigma^{>}(x, z)G^{<}(z, x) + \theta(z_0 - x_0)\Sigma^{<}(x, z)G^{<}(z, x) \\
&\theta(z_0 - x_0)G^{>}(x, z)\Sigma^{<}(z, x) - \theta(z_0 - x_0)G^{<}(x, z)\Sigma^{>}(z, x).
\end{aligned}
\tag{6.7}
$$

A simple calculation reveals that all four matrix elements are the same, so we have reduced our matrix of equations down to a single equation

$$
\begin{aligned}
&\lim_{x \to y}\left[K(x)\tilde{G}(x, y) - K(y)\tilde{G}(x, y)\right] \\
&= \int_{-\infty}^{x_0} \mathrm{d}z_0 \int \mathrm{d}^3z\left[\tilde{\Sigma}^{>}(x, z)\tilde{G}^{<}(z, x) - \tilde{G}^{<}(x, z)\tilde{\Sigma}^{>}(z, x)\right. \\
&\quad \left. - \tilde{\Sigma}^{<}(x, z)\tilde{G}^{>}(z, x) + \tilde{G}^{>}(x, z)\tilde{\Sigma}^{<}(z, x)\right].
\end{aligned}
\tag{6.8}
$$

Let us now consider the left-hand side of the above equation. We are free to choose any element, so let us choose the $G^{<}$ element. We can then show the left-hand side to be the divergence of the number current density

$$
\begin{aligned}
\mathrm{i}\lim_{x \to y}[\Box_x - \Box_y]G^{<}(x, y) &= \mathrm{i}\lim_{x \to y} \partial_\mu^x\left(\partial_x^\mu - \partial_y^\mu\right)G^{<}(x, y) \\
&= \lim_{x \to y} \mathrm{i}\partial_\mu^x\left(\partial_x^\mu - \partial_y^\mu\right)\left\langle \phi_-^\dagger(y)\phi_+(x)\right\rangle \\
&= \mathrm{i}\lim_{x \to y} \partial_\mu^x\left[\left\langle \phi_-^\dagger(y)\partial_x^\mu\phi_+(x)\right\rangle - \left\langle\left(\partial_y^\mu\phi_-^\dagger(y)\right)\phi_+(x)\right\rangle\right] \\
&= \mathrm{i}\partial_\mu\left\langle \phi_-^\dagger(x)\overleftrightarrow{\partial}^\mu\phi_+(x)\right\rangle = \partial^\mu j_\mu(x).
\end{aligned}
\tag{6.9}
$$

Putting the pieces together we have shown that the divergence of the current density is equal to a function of the self-energies

$$
\begin{aligned}
\partial_\mu j^\mu = &\int_{-\infty}^{x_0} \mathrm{d}z_0 \int \mathrm{d}^3z\left[\tilde{\Sigma}^{>}(x, z)\tilde{G}^{<}(z, x) - \tilde{G}^{<}(x, z)\tilde{\Sigma}^{>}(z, x)\right. \\
&\left. - \tilde{\Sigma}^{>}(x, z)\tilde{G}^{<}(z, x) + \tilde{G}^{>}(x, z)\tilde{\Sigma}^{<}(z, x)\right].
\end{aligned}
\tag{6.10}
$$

Note that the current divergences depend on an integral over the entire history of the system. This leads to the system being highly non-Markovian. Also note that the above has no time ordered or anti-time ordered propagators or functions, which are the propagators that appear in ordinary zero temperature QFT. Having the off-diagonal Green's functions be non-zero as you have in finite temperature is a necessary condition for the particle number current density to have a non-zero divergence. Repeating all the above steps for fermions except using the sum of the Schwinger–Dyson equations we obtain a similar equation

$$
\begin{aligned}
\partial_\mu j^\mu = &- \int_{-\infty}^{x_0} \mathrm{d}z_0 \int \mathrm{d}^3z \,\mathrm{Tr}\left[\tilde{\Sigma}^{>}(x, z)\tilde{S}^{<}(z, x) - \tilde{S}^{<}(x, z)\tilde{\Sigma}^{>}(z, x)\right. \\
&\left. - \tilde{\Sigma}^{>}(x, z)\tilde{S}^{<}(z, x) + \tilde{S}^{>}(x, z)\tilde{\Sigma}^{<}(z, x)\right].
\end{aligned}
\tag{6.11}
$$

To derive a network of coupled transport equations one then needs to systematically evaluate all self-energy contributions for each particle species.

6.2 Transport coefficients and sources

To systematically deal with each type of self-energy that contributes to the transport equations it is useful to categorize them. First and most important are interactions with the space–time varying vacuum. These are analogous to the zero temperature mass insertion diagrams and are often called vacuum expectation values (VEV) insertions. We make the so called 'VEV insertion' approximation where off-diagonal contributions to mass matrices are treated perturbatively. To calculate these diagrams we make the assumption that the mass basis is not changing rapidly during the phase transition. The strength of these interactions is controlled by the Yukawa/triscalar coupling which can carry a CP violating phase, in which case they become the dominant (tree-level) CP violating source. Additionally the strength of these sources is determined by the distance between the mass of the external and virtual state and the distance of both with the critical temperature. When the masses are nearly degenerate then there is a resonant boost to the CP violating sources [2–4, 7]. This is a particularly important feature as it implies that CP violating phases can be much smaller if one is near a resonance.

The VEV insertion diagrams also contain a CP conserving term which is also resonantly enhanced in strength when the masses of the virtual and external states are near degeneracy. This resonant enhancement of the relaxation term only slightly undermines the boost to the BAU produced via the resonant enhancement of the CP violating source.

All CP conserving interactions produce relaxation terms that ensure the number densities of the asymmetries, $n - \bar{n}$, all decay away from the bubble wall. Mathematically there is a qualitative difference between CP violating sources and relaxation terms. CP violating sources are functions of space–time coordinates independent of chemical potentials, while relaxation terms are proportional to the chemical potentials, which themselves are proportional to the number density. To see this consider the zeroth component of the particle number current. Sources have the form $\partial_t n(x) \sim f(x)$ and, if a source exists in the network of transport equations, the solution for n cannot be satisfied by $n(x) \equiv n(x) - \bar{n}(x) = 0$. In contrast, relaxation terms have the form $\partial_t n(x) \sim -\Gamma n(x)$, which has the trivial solution of $n(x) = 0$ in the absence of sources[2].

Apart from the VEV insertion diagrams, the other set of diagrams we consider that can be calculated perturbatively are the one-loop self-energies that involve Yukawa and triscalar interactions. These include supergauge interactions which, when sufficiently fast, are responsible for an approximate equilibrium between particles and their superpartners during the phase transition.

In order to tame the number of transport equations and interactions, one needs to consider which interactions are important. For a particle density to significantly

[2] Note that we sometimes refer to relaxation terms as CP conserving sources.

affect the dynamics of the combined left-handed number density ahead of the bubble wall, the particle must diffuse before the advancing wall catches up. For a diffusion constant D for a particular particle species the diffusion length is given by $\sqrt{Dt}$ while the distance the wall travels in the same time t is $v_{\mathrm{w}}t$. Taking $v_W = 0.05$ and $D = 50/T$ one has $t_{\mathrm{diff}} = 10^4/T$. Any process that occurs at a rate faster than the diffusion time is too significant to ignore, whereas if a particle couples to other particles only through rates slower than the diffusion time it can be approximately ignored in an analysis. Moreover, for a charge asymmetry to be transferred into an asymmetry in the left-handed number density, which seeds the ultimate baryon production by biasing the sphaleron processes, the particle number reactions involved in such a transfer must be fast enough that they occur before the advancing wall overtakes these particles. Incidentally, the electroweak sphaleron processes occur on a much larger time scale than the diffusion time. This means we can solve the dynamics that lead to an asymmetry in the left-handed number density first and use the calculated density as an input to a single differential equation for the production of the baryon number asymmetry.

Each self-energy is a formidable calculation and is sufficiently different from the sort of calculation one might encounter at zero temperature to warrant a significant degree of detailed explanation.

6.2.1 A biscalar with VEV insertions

Let us begin with the VEV insertions for scalar particles such as that given in figure 6.1. Here we can write the self-energy as $-g(x, y)G(x, y)$, where $g(x, y)$ contains all the information about the interaction with the space–time varying VEVs. The source equation is

$$\begin{aligned} J^\mu = &- \lim_{x \to z} \int \mathrm{d}^3z \int_{-\infty}^{x_0} \mathrm{d}z_0 \big[g(x, z)G_1^>(x, z)G_2^<(z, x) - g(z, x)G_1^>(x, z)G_2^<(z, x) \\ &+ g(z, x)G_1^<(x, z)G_2^>(z, x) - g(x, z)G_1^<(x, z)G_2^>(z, x) \big] \\ = &- \int \mathrm{d}^3z \int_{-\infty}^{x_0} \mathrm{d}z_0 g(x, z) \big[G_1^<(x, z)G_2^>(z, x) - G_1^>(x, z)G_2^<(z, x) \big]. \end{aligned} \tag{6.12}$$

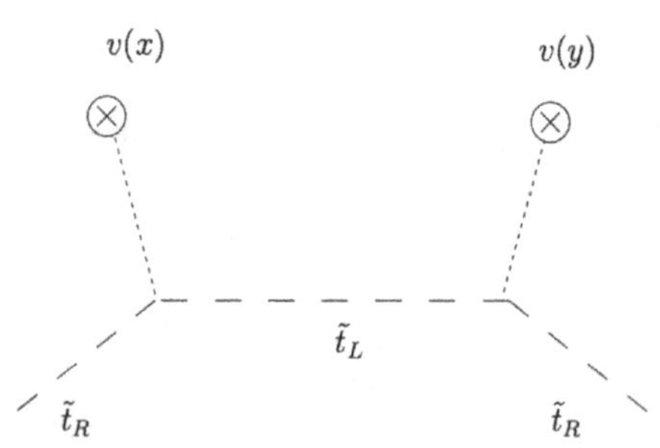

Figure 6.1. An example of a biscalar interaction with a space–time varying vacuum, also called a scalar VEV insertion. This particular diagram is relevant to any supersymmetric extension to the Standard Model.

It is possible to separate our source equation into CP conserving and violating terms Let us refer to the integrand as I. We can make use of the identities

$$\begin{aligned} I &= \frac{I}{2} + \frac{I^*}{2} + \frac{I}{2} - \frac{I^*}{2} \\ G(x, z) &= G^*(z, x) \end{aligned} \tag{6.13}$$

to show

$$\begin{aligned} S^{CP} = &- \lim x \to z \int \mathrm{d}^3 z \int_{-\infty}^{x_0} \mathrm{d}z_0 \{[g(x, z) + g(z, x)] \\ &\times \mathrm{Re}[G_1^>(x, z) G_2^<(z, x) - G_1^<(x, z) G_2^>(z, x)] + i[g(x, z) - g(z, x)] \\ &\times \mathrm{Im}[G_1^<(x, z) G_2^<(z, x) - G_1^<(x, z) G_2^>(z, x)]\} \\ \equiv &\, S^{CP} + S^{C/P}. \end{aligned} \tag{6.14}$$

We can see the need for CP violating phases as without them the second term is zero. A less trivial result is that we also see that we need a minimum of two scalar fields to create a tree-level CP violating source through VEV insertion diagrams, since we need $g(x, z) \neq g(z, x)$, which generally requires interference between interactions with each VEV. These VEV insertions are relatively large sources of CP violation and thus give a relatively large baryon asymmetry. It therefore takes some theoretical creativity to produce enough baryon asymmetry during the electroweak phase without any additional scalar field content to your model. Now let us proceed to a full calculation of the VEV insertion sources beginning with a few definitions as well as a few identities to make the calculation more efficient. Recall that our Green's functions are given by

$$G^\lambda(x, z) = \int \frac{\mathrm{d}^4 k}{(2\pi)^4} \mathrm{e}^{-ik\cdot(x-z)} g_B^\lambda(k_0, \mu_i) \rho(k_0, k), \tag{6.15}$$

where

$$\begin{aligned} \rho(k_0, k) &= \frac{\mathrm{i}}{2\omega_k} \left[\frac{1}{k_0 - \mathcal{E}} + \frac{1}{k_0 + \mathcal{E}} - \frac{1}{k_0 - \mathcal{E}^*} - \frac{1}{k_0 + \mathcal{E}^*} \right] \\ g_B^>(k_0, \mu_i) &= 1 + n_B(k_0 - \mu_i) \\ g_B^<(k_0, \mu_i) &= n_B(k_0 - \mu_i) \\ n_B(x) &= \frac{1}{\mathrm{e}^{\beta x} - 1}. \end{aligned} \tag{6.16}$$

In the above equations $\mathcal{E} = \omega + i\Gamma$ are the thermal widths and μ_i are the chemical potentials. We can then write the following useful identities

$$\begin{aligned} n_B(k_0 - \mu_i) &\approx n_B(k_0) - \mu_i \beta h_B(k_0) \\ g_B^>(-k_0, 0) &= - g_B^<(k_0, 0) \\ h(-k_0) &= h(k_0) \\ h_B(k_0) &= \frac{\mathrm{e}^{\beta k_0}}{(\mathrm{e}^{\beta k_0} - 1)^2}. \end{aligned} \tag{6.17}$$

Beginning with the CP conserving contribution to the source equation, let us expand S^{CP} to first-order in the chemical potentials, μ_i

$$S \approx S^0 + \delta S. \tag{6.18}$$

S^0 and δS both have poles from the function $\rho(k_0, k)$ at $\pm\mathcal{E}_i$ and $\pm\mathcal{E}_i^*$. We also have a tower of poles from the $n_B(k_0)$ and $h_B(k_0)$ functions at $k_0 = 2\pi ni/\beta$ for $n = 0, \pm1, \pm2\cdots$. We will later show that this tower of poles does not contribute to the source. For now let us include them and rewrite our source as

$$S^{CP} \approx (S^0 - A_1) + (\delta S - A_2) + A_1 + A_2, \tag{6.19}$$

where A_i contains the contributions from the poles at $2\pi ni/\beta$. We will also show that S^0 does not contribute. Let us begin with the non-zero term, $\delta S - A_2$ and take the limit of $x = z$. We can immediately infer that we have a delta function in the spatial components. It is also convenient to perform a change in variable $z^0 - x^0 \to z^0$. Taking the integral over the delta function we have

$$\begin{aligned}\delta S - A_2 = &- 2g(x, x)\beta \,\mathrm{Re} \int_{-\infty}^{0} \mathrm{d}z_0 \frac{\mathrm{d}^4k}{(2\pi)^4} \frac{\mathrm{d}q^0}{(2\pi)} e^{\mathrm{i}z^0(k^0-q^0)} \rho_1(k_0, k)\rho_2(q_0, k) \\ &\times \left[\mu_2 h_B(q_0) - \mu_1 h_B(k_0)\right].\end{aligned} \tag{6.20}$$

It is easy to lose track of factors of i, 2π and (-1). This becomes even more difficult with the one-loop self-energies. It is therefore useful to write the following rules for taking these contour integrals:

- a factor of (-1) for each '*';
- an overall factor of (-1) from the requirement that the contour over k^0 is in the opposite direction as the contour over q_0; and
- a factor of $2\pi\mathrm{i}^2/(2\omega_i)$ for each contour giving an overall factor of $(2\pi)^2/(4\omega_1\omega_2)$.

Using these rules to perform the contour integral the non-zero contribution to the source term is now

$$\begin{aligned}\delta S - A_2 = &+ 2g(x, x)\beta \,\mathrm{Re} \int_{-\infty}^{0} \frac{\mathrm{d}z_0}{4\omega_1\omega_2} \frac{\mathrm{d}^3k}{(2\pi)^3} (-1) e^{\mathrm{i}\left(\mathcal{E}_1^* - \mathcal{E}_2\right)z^0} \\ &\left[\mu_2 h_B(\mathcal{E}_2) - \mu_1 h_B\left(\mathcal{E}_1^*\right)\right] - e^{\mathrm{i}\left(-\mathcal{E}_1 + \mathcal{E}_2^*\right)z^0}\left[\mu_2 h_B(-\mathcal{E}_2) - \mu_1 h_B\left(-\mathcal{E}_1^*\right)\right] \\ &+ e^{\mathrm{i}\left(-\mathcal{E}_1 - \mathcal{E}_2\right)z^0}\left[\mu_2 h_B(\mathcal{E}_2) - \mu_1 h_B(-\mathcal{E}_1)\right] + e^{\mathrm{i}\left(\mathcal{E}_1^* + \mathcal{E}_2^*\right)z^0}\left[\mu_2 h_B\left(\mathcal{E}_2^*\right) - \mu_1 h_B\left(\mathcal{E}_1^*\right)\right].\end{aligned} \tag{6.21}$$

We obtain an additional factor of 4π when we integrate over the solid angle $\mathrm{d}\Omega$ and a factor of 2 from the real function. Finally the integral over z_0 is now straightforward. A simple calculation gives

$$\delta S - A_2 = \frac{g(x,x)\beta}{2\pi^2}\,\mathrm{Im}\int_0^\infty \frac{p\,\mathrm{d}p}{\omega_1\omega_2}\left[\frac{\mu_1 h_B(\mathcal{E}_1^*) - \mu_2 h_B(\mathcal{E}_2)}{\mathcal{E}_1^* - \mathcal{E}_2} + \frac{\mu_1 h_B(\mathcal{E}_1) - \mu_2 h_B(\mathcal{E}_2)}{\mathcal{E}_1 + \mathcal{E}_2}\right]. \tag{6.22}$$

Next we calculate $S^0 - A_1$. Since we only wish to show that the result is zero we can ignore all factors of proportionality

$$S^0 - A_1 \propto 2g(x,x)\beta\,\mathrm{Re}\int_{-\infty}^0 \mathrm{d}z_0 \frac{\mathrm{d}^4k}{(2\pi)^4}\frac{\mathrm{d}q^0}{(2\pi)} \\ e^{iz^0(k^0-q^0)}\rho_1(k_0,k)\rho_2(q_0,k)\left[g_B^>(k_0)g_B^<(q_0) - g_B^<(k_0)g_B^>(q_0)\right]. \tag{6.23}$$

Once again using our rules for taking these contour integrals which we defined before we can perform the contour integrals over k^0 and q^0 and write the integrand of the above equation, I_A, as

$$\begin{aligned} I_A = &(-1)e^{i(\mathcal{E}_1-\mathcal{E}_2^*)z^0}\left[g^>(\mathcal{E}_1)g^<(\mathcal{E}_2^*) - g^<(\mathcal{E}_1)g^>(\mathcal{E}_2^*)\right] \\ &+ e^{i(\mathcal{E}_1+\mathcal{E}_2)z^0}\left[g^>(\mathcal{E}_1)g^<(-\mathcal{E}_2) - g^<(\mathcal{E}_1)g^>(-\mathcal{E}_2)\right] \\ &+ e^{i(\mathcal{E}_1-\mathcal{E}_2^*)z^0}\left[g^>(-\mathcal{E}_1^*)g^<(\mathcal{E}_2^*) - g^<(-\mathcal{E}_1^*)g^>(\mathcal{E}_2^*)\right] \\ &+ (-1)e^{i(-\mathcal{E}_1^*+\mathcal{E}_2)z^0}\left[g^>(-\mathcal{E}_1^*)g^<(-\mathcal{E}_2) - g^<(-\mathcal{E}_1^*)g^>(-\mathcal{E}_2)\right] \\ \to &\frac{-1}{i(\mathcal{E}_1 - \mathcal{E}_2^*)}\left[g^>(\mathcal{E}_1)g^<(\mathcal{E}_2^*) - g^<(\mathcal{E}_1)g^>(\mathcal{E}_2^*)\right] \\ &+ \frac{1}{i(\mathcal{E}_1^* - \mathcal{E}_2)}\left[g^>(\mathcal{E}_1^*)g^<(\mathcal{E}_2) - g^<(\mathcal{E}_1^*)g^>(\mathcal{E}_2)\right] \\ = &\mathrm{Re}\left[\frac{A + A^*}{i}\right] = 0. \end{aligned} \tag{6.24}$$

So we have $S^0 = 0$. Finally for the A_i terms We need to write Bose distribution functions and its derivative as Laurent series, which means factoring out a factor of $1/x$ and $1/x^2$, respectively, and Taylor expanding the remaining piece

$$\begin{aligned} \frac{1}{e^{\beta x} - 1} &= \frac{1}{x}\frac{x}{e^{\beta x} - 1} = \frac{1}{x}\sum_{n=0}^{\infty} a_n x^n \\ h_B(x) &= \frac{1}{x^2}\frac{x^2 e^{\beta x}}{(e^{\beta x} - 1)^2} = \frac{1}{x^2}\sum_{n=0}^{\infty} b_n x^n. \end{aligned} \tag{6.25}$$

In both cases the sum is just a Taylor expansion with $a_0 = b_0 = 1/\beta$. For the $n \neq 0$ poles parallel to the real line, we can then just use the residue theorem and substitute the poles $p_0 = 2\pi in\beta$ into the rest of the integrand and find that the function $\rho(k_0, k) \to \frac{i}{2\omega_k}[\delta(\omega - i2\pi n) + \delta(\omega + i2\pi n)]$ when you take the thermal widths to zero. These delta functions cannot be satisfied over the range of the integral so they do not contribute. For the $n = 0$ poles these also do not contribute to the delta function as $\rho(k_0, k) = 0$ exactly when $k_0 = 0$ so this point does not contribute to the integral. Therefore we can write the CP conserving contribution to the source equation as just the non-zero term we calculated earlier, which was linear in the chemical potentials. We now explicitly write the function $g(x, z)$ for the MSSM VEV insertion diagram for stops as shown in figure 6.1

$$g(x, z) = y_t^2\left[A_t v_u(x) - \mu^* v_d(x)\right]\left[A_t^* v_u(x) - \mu v_d(x)\right]. \tag{6.26}$$

It is useful to write the source equation as

$$S^{CP} = \Gamma_{\tilde{t}}^{+}(\mu_1 + \mu_2) + \Gamma_{\tilde{t}}^{-}(\mu_1 - \mu_2), \tag{6.27}$$

with

$$\begin{aligned}\Gamma_m^{\pm} = &-\frac{\beta N_C y_t^2}{4\pi^2}|A_t v_u(x) - \mu^* v_d(x)|^2 \int_0^\infty \frac{dk k^2}{\omega_L \omega_R} \\ &\times \mathrm{Im}\left\{\frac{h_B(\mathcal{E}_L) \mp h_B(\mathcal{E}_R^*)}{\mathcal{E}_L - \mathcal{E}_R^*} - \frac{h_B(\mathcal{E}_L) \mp h_B(\mathcal{E}_R)}{\mathcal{E}_L + \mathcal{E}_R}\right\}.\end{aligned} \tag{6.28}$$

Next let us calculate the CP violating source. In this case the difference between the space–time-dependent functions in the limit of $z \to x$ takes the form

$$\lim_{z\to x}\left[g(x, z) - g(z, x)\right] = 2iy_t^2\,\mathrm{Im}(\mu A_t)\left[v_d(x)\partial_\lambda v_u(x) - \partial_\lambda v_d(x) v_u(x)\right](z - x)^\lambda. \tag{6.29}$$

Under the assumption of spatial isotropy only the time component of the above has a non-zero contribution, which is

$$\lim_{z\to x}\left[g(x, z) - g(z, x)\right] = 2iy_t^2\,\mathrm{Im}(\mu A_t)v^2(x)\dot{\beta}(x)(z - x)^0. \tag{6.30}$$

Once again the spatial integral gives a delta function in the three momenta. The $(z - x)^0$ factor is handled by integrating the time component by parts. Performing the contour integrals over k_0 and p_0, and once again expanding $n_B(k_0, \mu_i)$ in powers of μ_i, we find that the zeroth-order but not the first-order is non-zero. After some algebra one obtains

$$\begin{aligned}S_{\tilde{t}_R}^{C/P} = &\frac{N_C y_t^2}{2\pi^2}\mathrm{Im}[\mu A_t]v^2(x)\dot{\beta}\int_0^\infty \frac{dk k^2}{\omega_R \omega_L}\mathrm{Im}\left\{\frac{n_B(\mathcal{E}_R^*) - n_B(\mathcal{E}_L)}{(\mathcal{E}_L - \mathcal{E}_R^*)^2}\right. \\ &\left.+\frac{1 + n_B(\mathcal{E}_R) + n_B(\mathcal{E}_L)}{(\mathcal{E}_L + \mathcal{E}_R)^2}\right\}.\end{aligned} \tag{6.31}$$

Both the CP conserving relaxation term and the CP violating source acquire a resonant boost in their size when the masses of the particles are nearly degenerate. Defining

$$S_0^{C/P} = \int_0^\infty \frac{dkk^2}{\omega_R\omega_L} \mathrm{Im}\left\{ \frac{n_B(\mathcal{E}_R^*) - n_B(\mathcal{E}_L)}{\left(\mathcal{E}_L - \mathcal{E}_R^*\right)^2} + \frac{1 + n_B(\mathcal{E}_R) + n_B(\mathcal{E}_L)}{(\mathcal{E}_L + \mathcal{E}_R)^2} \right\} \tag{6.32}$$

and

$$\Gamma_0 = -\int_0^\infty \frac{dkk^2}{\omega_L\omega_R} \mathrm{Im}\left\{ \frac{h_B(\mathcal{E}_L) + h_B\left(\mathcal{E}_R^*\right)}{\mathcal{E}_L - \mathcal{E}_R^*} - \frac{h_B(\mathcal{E}_L) + h_B(\mathcal{E}_R)}{\mathcal{E}_L + \mathcal{E}_R} \right\} \tag{6.33}$$

we plot in figure 6.2 the value of S_0 and Γ_0 as a function of m_R for $T = m_L = 100$. The effect of the resonance is dramatic although there is some evidence that the center of the resonance may be severely damped when one goes beyond the approximations inherent in the above calculations [8]. This issue we return to in the discussion in chapter 12.

Before concluding this section we mention the scope of the approximation schemes we have used. If the mass of the virtual state is more than a factor of ~2 different to the mass of the in-state then the CP conserving relaxation term does not go to zero but actually becomes negative, which spuriously implies that baryon asymmetry can be generated even in the absence of a CP violating phase. To deal with this one needs to sanitize both the CP violating source and the CP conserving relaxation terms by setting them to zero by hand for a large mass splitting. The cutoff where this happens depends on the thermal widths. This is a generic feature of all VEV insertion diagrams, including the ones we will consider in the subsequent subsections.

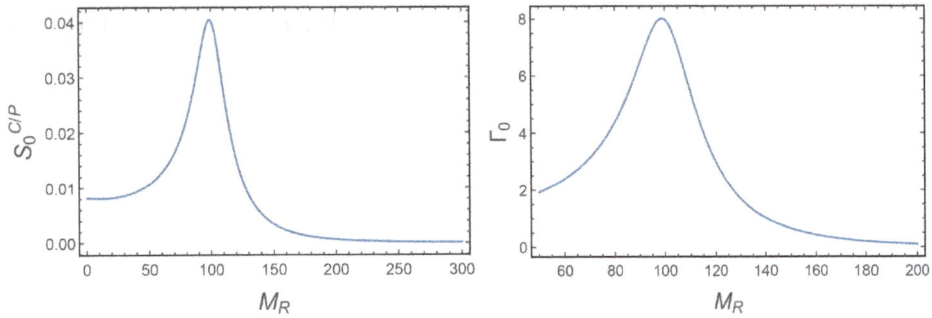

Figure 6.2. Plot of CPV source and CP conserving relaxation terms involving biscalar interaction with a space–time varying vacuum as a function of one of the masses. The above figures show the resonance for degenerate masses where $M_L = 100$, $T = 100$ and $\Gamma_L = \Gamma_R = 5$.

6.2.2 Dirac fermions with VEV insertions

For an example of an interaction involving Dirac fermions contributing significantly to the baryon asymmetry, consider the Higgsino and gaugino interacting with the space–time varying vacuum in the MSSM. First write the Higgsino mass terms in the mass basis of the symmetric phase

$$L_{\tilde{H}} \ni |\mu|\left(\tilde{H}_d^0\tilde{H}_u^0 - \tilde{H}_d^-\tilde{H}_u^+ + \tilde{H}_d^{0\dagger}\tilde{H}_u^{0\dagger} - \tilde{H}_d^{-\dagger}\tilde{H}_u^{+\dagger}\right). \tag{6.34}$$

Even though the Higgsino is a Majorana fermion, we can define three Dirac spinors out of the charged and neutral Higgsinos as well as a third out of the wino

$$\Psi_{\tilde{H}^+} = \begin{pmatrix} \tilde{H}_u^+ \\ \tilde{H}_d^{-\dagger} \end{pmatrix} \Psi_{\tilde{H}^0} = \begin{pmatrix} -\tilde{H}_u^0 \\ \tilde{H}_d^{0\dagger} \end{pmatrix} \Psi_{\tilde{W}^\pm} = \begin{pmatrix} \tilde{W}^+ \\ \tilde{W}^- \end{pmatrix}. \tag{6.35}$$

These fermions all obey the Dirac equation with masses ($|\mu|$, $|\mu|$, M_2), respectively. This means we can define a vector charge for these particle species. There are two gauginos that interact with the space–time varying vacuum that are generally ignored, the bino and the third wino, $\tilde{W}^3$. The justification for this is partly because one cannot define a vector charge for these particle species, which complicates matters, and partly because their contribution has been shown to be small [4]. The terms left in the Lagrangian that involve gaugino–Higgsino–vacuum interactions are

$$\begin{aligned} L_{\text{int}} &\ni - g_2\bar{\Psi}_{\tilde{H}}\left[v_d(x)P_L + v_u(x)e^{i\theta_\mu}P_R\right]\Psi_{\tilde{W}^+} \\ &\equiv - g_2\bar{\Psi}_{\tilde{H}}\left[g_L(x)P_L + g_R(x)P_R\right]\Psi_{\tilde{W}^+}\Big] + h.\,c., \end{aligned} \tag{6.36}$$

where in the second line we have implicitly defined $g_{L/R}$ for notational convenience. The self-energy for interactions of the form shown in figure 6.3 is

$$\Sigma_{\tilde{H}^\pm}(x, z) = -g_2^2\left[g_L(x)P_L + g_R(x)P_R\right]S_{\tilde{W}^\pm}(x, z)\left[g_L(z)^*P_R + g_R(z)^*P_L\right]. \tag{6.37}$$

We then want to write the divergence of the current that arises from the above self-energy as a sum of a CP conserving relaxation term and a CP violating source term, just like we did before. However, this takes a little more care than before as the

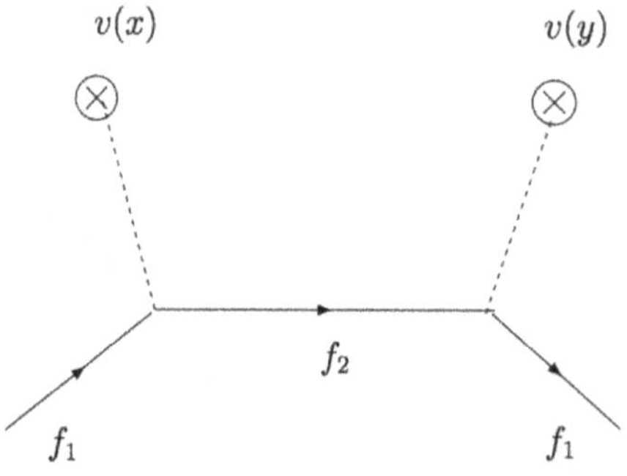

Figure 6.3. Depiction of a fermion interacting with a space–time varying vacuum via VEV insertion.

self-energy contains projection operators and the propagators contain Dirac matrices. Let us define

$$S^{\lambda}(k) = \bar{S}^{\lambda}(k)(\not{k} + m) \tag{6.38}$$

to allow us to separate our treatment of the matrices and the functions. When we calculate terms such as $\sim\int \Sigma^{>}_{\tilde{H}^{\pm}}(x, z)S^{<}_{\tilde{W}^{\pm}}(z, x) + \cdots$ we will encounter terms like the following (omitting measures and other details that detract from explaining the step):

$$\begin{aligned}
&\int \mathrm{Tr}\,[g_L(x)P_L + g_R(x)P_R]\bar{S}_{\tilde{W}}(k)(\not{k} + |\mu|) \\
&\quad\times\left[g_L^*(z)P_R + g_R^*(z)P_L\right]\bar{S}_{\tilde{H}}(p)\left(\not{p} + M_2\right) \\
&\quad= \int \mathrm{Tr}\left\{\left[g_L(x)g_R^*(z)P_L + g_R(x)g_L(z)^*P_R\right]\not{k}\,\not{p}\right. \\
&\quad+ \left.M_2|\mu|\left[g_L(x)g_R(z)^*P_L + g_R(x)g_L(z)^*P_R\right]\right\} \\
&\quad\times \bar{S}^{>}_{\tilde{W}}(k)\bar{S}^{<}_{\tilde{H}}(p)\mathrm{e}^{-\mathrm{i}(x-z)\cdot(k-p)}.
\end{aligned} \tag{6.39}$$

We can then define the space–time functions that multiply the $\not{k}\,\not{p}$ terms and the mass terms

$$\begin{aligned}
g_A(x, z) &= \frac{g_2^2}{2}\left[g_L(x)g_L(z)^* + g_R(x)g_R(z)^*\right] \\
g_B(x, z) &= \frac{g_2^2}{2}\left[g_L(x)g_R(z)^* + g_R(x)g_L(z)^*\right]
\end{aligned} \tag{6.40}$$

and note we have the identity that $g_j(x, z)^* = g(z, x)$ to write the current divergence for the gaugino–Higgsino–vacuum as a sum of a CP conserving relaxation term and a CP violating source term as before. It can be written in a compact form using the notation that $\mathrm{Tr}[\cdot]_A$ and $\mathrm{Tr}[\cdot]_B$ correspond to the terms involving the $\not{k}\,\not{p}$ piece and the mass piece, respectively,

$$\begin{aligned}
S_{\tilde{H}^{\pm}}(x) &= \int \mathrm{d}^3z \int_{-\infty}^{0} \mathrm{d}z_0 \sum_{j\in(A,B)} \left[g_j(x, z) + g_j(z, x)\right] \\
&\quad\times \mathrm{ReTr}\left[S^{>}_{\tilde{W}^{\pm}}(x, z)S^{<}_{\tilde{H}^{\pm}}(z, x) - S^{<}_{\tilde{W}^{\pm}}(x, z)S^{>}_{\tilde{H}^{\pm}}(z, x)\right]_j \\
&= i\left[g_j(x, z) - g_j(z, x)\right]\mathrm{ImTr}\left[S^{>}_{\tilde{W}^{\pm}}(x, z)S^{<}_{\tilde{H}^{\pm}}(z, x)\right. \\
&\quad\left.-S^{<}_{\tilde{W}^{\pm}}(x, z)S^{>}_{\tilde{H}^{\pm}}(z, x)\right]_j.
\end{aligned} \tag{6.41}$$

To solve these integrals, the steps are very similar to the biscalar–vacuum interaction so we just quote the result. The CP conserving term is given by

$$S_{\tilde{H}}^{\not{CP}}(x) = \Gamma^{+}_{\tilde{H}^{\pm}}(\mu_{\tilde{W}^{+}} + \mu_{\tilde{H}^{+}}) + \Gamma^{-}_{\tilde{H}^{\pm}}(\mu_{\tilde{W}^{+}} - \mu_{\tilde{H}^{+}}) \tag{6.42}$$

with

$$\Gamma^{\pm}_{\tilde{H}^{\pm}} = \frac{1}{T}\frac{g_2^2}{2\pi^2}v(x)^2\int_0^{\infty}\frac{dkk^2}{\omega_{\tilde{H}}\omega_{\tilde{W}}}\mathrm{Im}\Big\{\Big[\mathcal{E}_{\tilde{W}}\mathcal{E}^*_{\tilde{H}} - k^2M_2|\mu|\cos\theta_\mu\sin 2\beta\Big]\frac{h_F(\mathcal{E}_{\tilde{W}}) \mp h_F\big(\mathcal{E}^*_{\tilde{H}}\big)}{\mathcal{E}_{\tilde{W}} - \mathcal{E}^*_{\tilde{H}}} + \Big[\mathcal{E}_{\tilde{W}}\mathcal{E}_{\tilde{H}} + k^2 - M_2|\mu|\cos\theta_\mu\sin 2\beta\Big]\frac{h_F(\mathcal{E}_{\tilde{W}}) \mp h_F(\mathcal{E}_{\tilde{H}})}{\mathcal{E}_{\tilde{W}} + \mathcal{E}_{\tilde{H}}}\Big\} \quad (6.43)$$

and the function $h_F(x) = e^{x/T}/(e^{x/T}+1)^2$ defined analogously to the bosonic case. It is straightforward to derive the CP conserving term for the other Dirac fermions. If one follows the above derivation for $\tilde{H}^0$ states, one will find that we merely make the replacement $g_2 \to g_2/\sqrt{2}$ when the virtual state is the wino. There is also a term for when the external state is the wino and the virtual state is the bino, which requires the frequency and the thermal widths to be changed appropriately (i.e. $\omega_{\tilde{W}} \to \omega_{\tilde{B}}$ and $\Gamma_{\tilde{W}} \to \Gamma_{\tilde{B}}$). This is similarly true for the *CP* violating sources,

$$S^{CP}_{\tilde{H}^{\pm}} = \frac{g_2^2}{\pi^2}v(x)^2\dot{\beta}(x)M_2|\mu|\sin\theta_\mu \times \int_0^{\infty}\frac{dkk^2}{\omega_{\tilde{H}}\omega_{\tilde{W}}}\mathrm{Im}\left\{\frac{n_F(\mathcal{E}_{\tilde{W}}) - n_F\big(\mathcal{E}^*_{\tilde{H}}\big)}{(\mathcal{E}_{\tilde{W}} - \mathcal{E}_{\tilde{H}^*})^2} + \frac{1 - n_F(\mathcal{E}_{\tilde{W}}) - n_F(\mathcal{E}_{\tilde{H}})}{(\mathcal{E}_{\tilde{W}} + \mathcal{E}_{\tilde{H}})^2}\right\}. \quad (6.44)$$

Just like the biscalar interactions, these terms have a resonant boost near where the mass is degenerate. Also the approximation scheme used breaks down when the masses are very non-degenerate or exactly degenerate.

6.2.3 Chiral fermions with VEV insertions

The final VEV insertion term we will consider is the interaction between the top quarks and vacuum. There only arises a single CP conserving relaxation term. Interactions between the top quark and multi-particle states lead to additional poles of the propagator whose contribution is too large to ignore [9–10]. For a massless top quark the top quark propagator in the presence of a thermal bath becomes

$$S^{\lambda}(x, y, \mu_i) = \int\frac{d^4k}{(2\pi)^4}e^{-ik\cdot(x-y)}g_F^{\lambda}(k_0, \mu) \times\left[\frac{\gamma_0 - \vec{\gamma}\cdot\hat{k}}{}\rho_+\big(k_0, \vec{k}, \mu_i\big) + \frac{\gamma_0 + \vec{\gamma}\cdot\hat{k}}{2}\rho_-\big(k_0, \vec{k}, \mu_i\big)\right] \quad (6.45)$$

where the spectral densities are related to each other via $\rho_-(k_0, \vec{k}, \mu_i)=[\rho_+(-k^{0*}, \vec{k}, -\mu_i)]^*$ and are defined by

$$\rho_+\left(k_0, \vec{k}, \mu_i\right) = i\left[\frac{Z_p(k, \mu_i)}{k_0 - \mathcal{E}_p(k, \mu_i)} - \frac{Z_p(k, \mu_i)^*}{k_0 - \mathcal{E}_p(k, \mu_i)^*} + \frac{Z_h(k, -\mu_i)}{k_0 + \mathcal{E}_h(k, -\mu_i)^*} - \frac{Z_h(k, -\mu_i)}{k_0 + \mathcal{E}_h(k, -\mu_i)} + \cdots\right]. \tag{6.46}$$

Here the dots denote the non-pole contributions which we neglect. In the above equation, $Z_{p,h}$ are related to the poles of the equation

$$0 = k^0 - k - \frac{\alpha_s C_F \pi T^2}{2}\left[\left(1 - \frac{k^0}{k}\right)\log\left|\frac{k^0 + k}{k^0 - k}\right| + 2\right]. \tag{6.47}$$

Explicitly, if we define the poles to the above equation by $k^0_{\text{pole}} = \{E_p(k), -E_h(k)\}$, then we have

$$Z_{p,\,h} = \frac{E^2_{p,\,h} - k^2}{m_f^2} \tag{6.48}$$

and the thermal mass (see section 7.4) is $m_f^2 = \alpha_s C_F \pi T^2/2$ where C_F is the Casimir of the group. There are two thermal widths $\Gamma_{p,h}$, which shift the poles in the imaginary direction as before,

$$\mathcal{E}_x = \sqrt{k^2 + Z_x(k, m)^2 m^2} - \mathrm{i}\Gamma_x. \tag{6.49}$$

It is useful to write approximations for Z_p and Z_h from [9]. In the $k \ll m$ regime, we have the approximations

$$Z_p(k, m) \approx \frac{1}{2} + \frac{k}{6m} \tag{6.50}$$

$$Z_h(k, m) \approx \frac{1}{2} - \frac{k}{6m} \tag{6.51}$$

and the following in the regime $k \gg m$

$$Z_p(k, m) \approx 1 + \left(\frac{m^2}{2k^2}\right)\left(1 - \log\left[2\frac{k^2}{m^2}\right]\right) \tag{6.52}$$

$$Z_h(k, m) \approx \left(\frac{2k^2}{em^2}\right)e^{-2\frac{k^2}{m^2}}. \tag{6.53}$$

In the intermediate region we take a k/m dependent weighted average of the functions. Using interacting fermion propagators one can derive the rate

$$\Gamma^{\pm}_{m,t_R} = \frac{1}{T}\frac{N_C y_t^2 v^2(x)\sin^2\beta}{\pi^2}\int_0^\infty dk k^2\,\mathrm{Im}\left\{\frac{Z_p^R(k)Z_p^L(k)}{\mathcal{E}_p^R + \mathcal{E}_p^L}\left[h_F(\mathcal{E}_p^L) \mp h_F(\mathcal{E}_p^R)\right]\right. \\ \left. + \frac{Z_h^R(k)^* Z_p^L(k)}{\mathcal{E}_p^L - \mathcal{E}_h^{R*}}\left[h_F(\mathcal{E}_p^L) \mp h_F(\mathcal{E}_h^{R*})\right] + (p \leftrightarrow h)\right\}. \tag{6.54}$$

6.2.4 Triscalar and Yukawa interactions

Let us now turn to the one-loop effects given in figure 6.4. Little new is learned from Yukawa and triscalar interactions so we will just look at the triscalar case and leave the Yukawa case as an exercise for the reader, merely quoting the result at the end. In the MSSM the most important such triscalar interaction—in the sense that it has the largest effect—is between the stops and the Higgs. There are three different diagrams with either the left- or right-handed stop or the Higgs as the 'in'-state. The three interactions will turn out to have a simple relation between them so we can just consider one. Choosing the right-handed stop as the in-state, the self-energy can be written as just a product of propagators and the coupling

$$\Sigma^\lambda = |\lambda|^2 G_L^\lambda G_H^\lambda. \tag{6.55}$$

Here we have written the stop–Higgs triscalar coupling as λ to make the generalization clear. We will use the approximation of setting the thermal widths to be infinitesimal. Also in our analysis the following identity will be useful:

$$g^>(-\omega, \mu)\big|_{\mu=0} = -g^<(\omega, \mu)\big|_{\mu=0}, \tag{6.56}$$

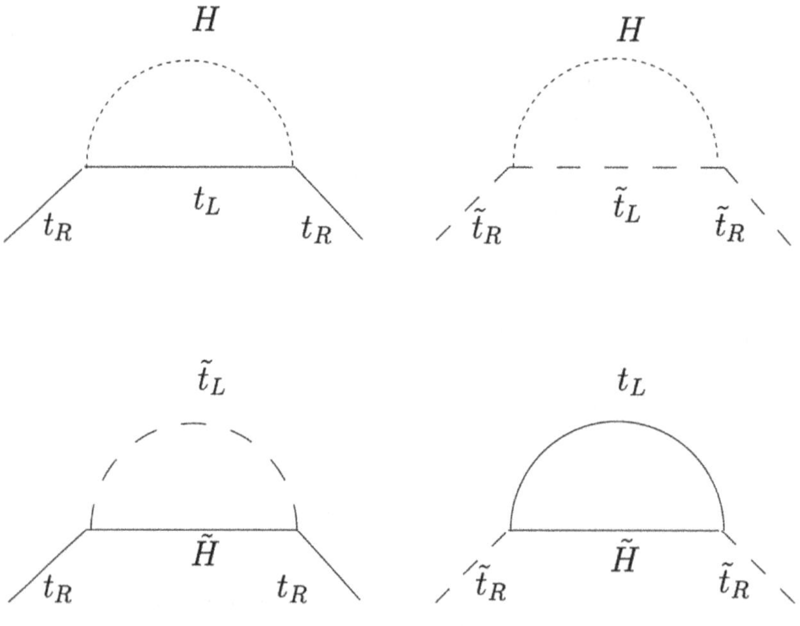

Figure 6.4. Most relevant one loop MSSM Yukawa and triscalar interactions involving tops, stops, Higgs, and Higgsinos. Other triscalar and Yukawa interactions are suppressed by small Yukawa couplings. Reproduced from [7].

which holds for both fermions and bosons. From here on we will denote g^λ at zero μ simply as $g^\lambda(\omega)$. This identity is of course on top of the previous dictionary of identities introduced in the biscalar section. With this in mind we can write to first-order in μ that $\pm g^<_{B,F}(\omega, \mu) \approx g^<_{B,F}(\omega) - \beta\mu h_{B,F}(\omega)$, where $h_{B,F}(\omega) \equiv \exp(\beta\omega)/\{[\exp(\beta\omega) \mp 1]^2\}$. There is no CP violating source that arises from this self-energy, only a CP conserving relaxation term. Therefore there cannot be a contribution from terms that are zeroth-order in the chemical potentials, as such a term would cause a particle–anti-particle asymmetry even in the absence of CP violation. Therefore we can expand the propagators, G, and by implication the source S, to first-order in the chemical potentials, μ_i. That is, $G \approx G^0 + \delta G$. We once again separate the terms that relate to the residues of the poles at $\mu_i + 2\pi n i/\beta = \{k_0, p_0, q_0\}$ and denote the sum of all these terms as 'A'. These terms we will ignore as they will not contribute. Our source to first-order in the chemical potentials is now

$$S = (S^0 - A) + (\delta S - A) + 2A. \tag{6.57}$$

It can easily be shown mathematically that $S = \delta S$ as all other terms are zero. The proof of this, however, is very analogous to the calculation in the biscalar VEV insertion case, albeit messier and little more is learned from this. Since we have intuition as to why S^0 and A are zero we will omit this calculation. Using the identity that $\delta G^< = \delta G^>$ there are three terms for each product of the Green's functions leading to an expansion involving 12 terms in total

$$\begin{aligned}\delta S = |\lambda|^2 \int d^3z \int_{-\infty}^{X_0} dz_0 \Big[&\delta G_L(x, z) G^>_H(x, z) G^<_R(z, x) \\ &+ G^>_L(x, z)\delta G_H(x, z) G^<_R(z, x) + G^>_L(x, z) H^>(x, z)\delta G_R(z, x) + \cdots \Big].\end{aligned} \tag{6.58}$$

We can factorize the expansion as the identity that $\delta G^> = \delta G^<$ introduces some degeneracies

$$\begin{aligned}\delta S = |\lambda|^2 \int d^3z \int_{-\infty}^{X_0} & dz_0 \delta G_L(x, z)\Big[G^>_H(x, z) G_R < (z, x) - G^<_H(x, z) G^>_R(z, x)\Big] \\ &+ \delta G_L(z, x)\Big[G^>_H(z, x) G_R < (x, z) - G^<_H(z, x) G^>_R(x, z)\Big] \\ &+ \delta G_H(x, z)\Big[G^>_L(x, z) G^<_R(z, x) - G^<_L(x, z) G^>_R(z, x)\Big] \\ &+ \delta G_H(z, x)\Big[G^>_L(z, x) G^<_R(x, z) - G^<_L(z, x) G^>_R(x, z)\Big] \\ &+ \delta G_R(x, z)\Big[G^>_L(x, z) G^>_H(z, x) - G^<_L(x, z) G^<_H(z, x)\Big] \\ &+ \delta G_R(z, x)\Big[G^>_L(z, x) G^>_H(x, z) - G^<_L(z, x) G^<_H(x, z)\Big] \\ \equiv\ & \delta S_L + \delta S_H + \delta S_R.\end{aligned} \tag{6.59}$$

For now let us focus on δS_L. We can substitute in the expressions for the scalar propagators, $G^\lambda(x, z)$, and use the identity $G^\lambda(x, y) = G^{\lambda *}(z, x)$ to write

$$\delta S_L = 2|\lambda|^2 \beta\mu_L \mathrm{Re}\left[\int \mathrm{d}^3 z \int_{-\infty}^{x_0} \mathrm{d}z_0 \frac{\mathrm{d}^4 k}{(2\pi)^4}\frac{\mathrm{d}^4 p}{(2\pi)^4}\frac{\mathrm{d}^4 q}{(2\pi)^4} \mathrm{e}^{-\mathrm{i}(k+p-q)(x-z)} h(k_0) \right.$$
$$\left. \times \rho(k_0, k)\rho(q_0, q)\rho(p_0, p)\left[g^>(p_0)g^<(q_0) - g^<(p_0)g^>(q_0)\right]\right]. \tag{6.60}$$

Performing the integrals over $\mathrm{d}^3 z$ makes a δ function for $k + p - q$. We then perform contour integrals for p_0 and q_0 omitting for now the overall sign of each term until we derive some rules for it. The contour over p_0 is carried out in the upper half of the Argand plane which contains the poles $\mathcal{E}_H^*$ and $-\mathcal{E}_H$. Similarly the contour for q_0 is carried out over the lower half of the complex plane covering poles $\mathcal{E}_R$ and $-\mathcal{E}_R^*$. We also change variables to $\tau = z_0 - x_0$ and rename τ as z_0. We then have

$$\delta S_L = 2|\lambda|^2 \beta\mu_L \,\mathrm{Re}\left[\int_{-\infty}^{0} \mathrm{d}z_0 \frac{\mathrm{d}k_0}{(2\pi)}\frac{\mathrm{d}^3 p}{(2\pi)^3}\frac{\mathrm{d}^3 q}{(2\pi)^3}\frac{1}{4\omega_R\omega_L} h(k_0)\rho_L(k_0, q-p)\right.$$
$$\times\left\{(-1)^{L_1}\times \mathrm{e}^{-\mathrm{i}\left(k_0+\mathcal{E}_H^*-\mathcal{E}_R\right)z_0}\left[g^>\left(\mathcal{E}_H^*\right)g^<(\mathcal{E}_R) - g^<\left(\mathcal{E}_H^*\right)g^>(\mathcal{E}_R)\right]\right.$$
$$+(-1)^{L_2}\times \mathrm{e}^{-\mathrm{i}\left(k_0+\mathcal{E}_H^*+\mathcal{E}_R^*\right)z_0}\left[g^>\left(\mathcal{E}_H^*\right)g^<\left(-\mathcal{E}_R^*\right) - g^<\left(\mathcal{E}_H\right)^* g^>\left(-\mathcal{E}_R^*\right)\right] \tag{6.61}$$
$$+(-1)^{L_3}\times \mathrm{e}^{-\mathrm{i}\left(k_0-\mathcal{E}_H-\mathcal{E}_R\right)z_0}\left[g^>(-\mathcal{E}_H)g^<(\mathcal{E}_R) - g^<(-\mathcal{E}_H)g^>(\mathcal{E}_R)\right]$$
$$\left.\left.+(-1)^{L_1}\times \mathrm{e}^{-\mathrm{i}\left(k_0-\mathcal{E}_H+\mathcal{E}_R^*\right)z_0}\left[g^>(-\mathcal{E}_H)g^<\left(-\mathcal{E}_R^*\right) - g^<(-\mathcal{E}_H)g^>\left(-\mathcal{E}_R^*\right)\right]\right\}\right].$$

To calculate the signs, L_i, we note that we acquire a factor of (-1) for each term from the contour integration itself as well as a factor of (-1) for every instance of a $\mathcal{E}^*$ term. Using these rules we can derive $L_1 = 2$, $L_2 = 3$, $L_3 = 1$, and $L_4 = 2$. Exploiting the presence of the Re[·] function we can turn the integral over z_0 into a δ function and use the identity $g^>(-x) = -g^<(x)$ to write

$$\delta S_L = 2\beta\mu_L \frac{|\lambda|^2}{2}\mathrm{Re}\left[\int \frac{\mathrm{d}^3 p}{(2\pi)^3}\frac{\mathrm{d}^3 q}{(2\pi)^3}\frac{1}{4\omega_R\omega_H}\right.$$
$$\times\left\{h\left(\mathcal{E}_R - \mathcal{E}_H^*\right)\rho_L\left(\mathcal{E}_R - \mathcal{E}_H^*, q-p\right)\left[g^>\left(\mathcal{E}_H^*\right)g^<(\mathcal{E}_R) - g^<\left(\mathcal{E}_H^*\right)g^>(\mathcal{E}_R)\right]\right.$$
$$- h\left(-\mathcal{E}_R^* - \mathcal{E}_H^*\right)\rho_L\left(-\mathcal{E}_R^* - \mathcal{E}_H^*, q-p\right)$$
$$\times\left[-g^>\left(\mathcal{E}_H^*\right)g^>\left(\mathcal{E}_R^*\right) + g^<\left(\mathcal{E}_H^*\right)g^<\left(\mathcal{E}_R^*\right)\right] \tag{6.62}$$
$$- h(\mathcal{E}_R + \mathcal{E}_H)\rho_L(\mathcal{E}_R + \mathcal{E}_H, q-p)\left[g^<(\mathcal{E}_H)g^<(\mathcal{E}_R) - g^>(\mathcal{E}_H)g^>(\mathcal{E}_R)\right]$$
$$+ h\left(-\mathcal{E}_R^* + \mathcal{E}_H\right)\rho_L\left(-\mathcal{E}_R^* + \mathcal{E}_H, q-p\right)$$
$$\left.\left.\times\left[g^<\left(\mathcal{E}_H^*\right)g^>\left(\mathcal{E}_R^*\right) - g^>(\mathcal{E}_H)g^<\left(\mathcal{E}_R^*\right)\right]\right\}\right].$$

We now use the identities $\rho(x, y) = \rho^*(x, y)$, $\text{Re}[A + A^*] = \text{Re}[2a]$, $h(x) = h(-x)$ and $\rho(x, y) = -\rho(-x, y)$ which dramatically simplifies the form of the source

$$\begin{aligned}\delta S_L = 2|\lambda|^2 \beta\mu_L \,\text{Re}\Bigg[\int \frac{d^3p d^3q}{(2\pi)^6 4\omega_R\omega_H}\\ h\left(\mathcal{E}_R - \mathcal{E}_H^*\right)\rho\left(\mathcal{E}_R - \mathcal{E}_H^*, q - p\right)\left[g^>(\mathcal{E}_H^*)g^<(\mathcal{E}_R) - g^<(\mathcal{E}_H^*)g^>(\mathcal{E}_R)\right]\\ - h(\mathcal{E}_H + \mathcal{E}_R)\rho(\mathcal{E}_H + \mathcal{E}_R, q - p)[g^>(\mathcal{E}_H)g^>(\mathcal{E}_R) - g^<(\mathcal{E}_H)g^<(\mathcal{E}_R)]\Bigg].\end{aligned} \tag{6.63}$$

The integrals over the angles can be performed by defining the angle between p and q such that $p \cdot q = pqx$. The rest of the angular integrals just produce a factor of $2(2\pi)^2$. We then perform a change of variables $\omega \equiv \omega_H + \sqrt{q^2 - p^2 - 2pqx + M_L^2}$ with new limits $\omega \pm \equiv \omega_H + \sqrt{q^2 - p^2 \pm 2pq + M_L^2}$ (chosen to absorb an extra minus sign) which allows us to write the more tractable integral

$$\begin{aligned}\delta S = |\lambda|^2 \beta\mu_L \,\text{Re}\Bigg[\int_{\omega-}^{\omega^+} d\omega \frac{(pq)dpdq}{(2\pi)^4 2\omega_R\omega_H}\\ \times h\left(\mathcal{E}_R - \mathcal{E}_H^*\right)\rho'\left(\mathcal{E}_R - \mathcal{E}_H^*, q - p\right)\left[g^>(\mathcal{E}_H^*)g^<(\mathcal{E}_R) - g^<(\mathcal{E}_H^*)g^>(\mathcal{E}_R)\right]\\ - h(\mathcal{E}_H + \mathcal{E}_R)\rho'(\mathcal{E}_H + \mathcal{E}_R, q - p)\left[g^>(\mathcal{E}_H)g^>(\mathcal{E}_R) - g^<(\mathcal{E}_H)g^<(\mathcal{E}_R)\right]\Bigg],\end{aligned} \tag{6.64}$$

where $\rho' \equiv \rho \times 2\omega$. Let us further simplify the above integral by performing manipulations on the following part of the integrand

$$\begin{aligned} h(\mathcal{E}_R - \mathcal{E}_H^*)&[g^>(\mathcal{E}_H^*)g^<(\mathcal{E}_R) - g^<(\mathcal{E}_H^*)g^>(\mathcal{E}_R)]\\ &= \frac{\exp[\beta(\mathcal{E}_R - \mathcal{E}_H^*)]}{\left\{\exp\left[\beta\left(\mathcal{E}_r - \mathcal{E}_H^*\right)\right] - 1\right\}^2}\left[n_B(\mathcal{E}_R) - n_B\left(\mathcal{E}_H^*\right)\right]\\ &= \frac{\exp[\beta(\mathcal{E}_R - \mathcal{E}_H^*)]}{\left\{\exp\left[\beta\left(\mathcal{E}_r - \mathcal{E}_H^*\right)\right] - 1\right\}^2}\left[\frac{\exp\beta\mathcal{E}_H^* - \exp\beta\mathcal{E}_R}{\left(\exp\beta\mathcal{E}_H^* 1\right)\left(\exp\beta\mathcal{E}_R - 1\right)}\right]\\ &= -\frac{\exp\beta\mathcal{E}_R}{\left(\exp\left[\beta\left(\mathcal{E}_R - \mathcal{E}_H^*\right)\right] - 1\right)\left(\exp\beta\mathcal{E}_R - 1\right)\left(\exp\beta\mathcal{E}_H^* - 1\right)} \equiv I_L.\end{aligned} \tag{6.65}$$

We can also derive the analogous terms I_H and I_R to be

$$\begin{aligned} I_H &= h(\mathcal{E}_H^*)[n_B(\mathcal{E}_R) - n_B(\mathcal{E}_H^*)]\\ &= -\frac{\exp\beta\mathcal{E}_R}{(\exp[\beta(\mathcal{E}_R - \mathcal{E}_H^*)] - 1)(\exp\beta\mathcal{E}_R - 1)(\exp\beta\mathcal{E}_H^* - 1)}\end{aligned} \tag{6.66}$$

and

$$\begin{aligned} I_R &= h(\mathcal{E}_R)[1 + n_B(\mathcal{E}_R - \mathcal{E}_H^*) + n_B(\mathcal{E}_H^*)]\\ &= +\frac{\exp\beta\mathcal{E}_R}{(\exp[\beta(\mathcal{E}_R - \mathcal{E}_H^*)] - 1)(\exp\beta\mathcal{E}_R - 1)(\exp\beta\mathcal{E}_H^* - 1)},\end{aligned} \tag{6.67}$$

respectively. The other terms in the integrand are nothing more than the first term with $\mathcal{E}_H^* \rightarrow -\mathcal{E}_H$ and can be similarly simplified without putting pen to paper. Finally we can make a change of variables from the respective momentum variables to frequencies. Recalling that $\delta S = S$, which we will show at the end of this calculation, our expression for the source term S_L is then [11]

$$S_L = (\mu_R - \mu_L - \mu_H)\Gamma_Y, \tag{6.68}$$

where

$$\begin{aligned}\Gamma_Y = |\lambda|^2\beta\mathrm{Re}\Bigg[\int_{m_R,m_H}^{\infty} \frac{\mathrm{d}\omega_R\mathrm{d}\omega_H}{2(2\pi)^4}\int_{\omega-}^{\omega_+}\mathrm{d}\omega \\ \Big\{h\big(\mathcal{E}_R - \mathcal{E}_H^*\big)\rho_L'\big(\mathcal{E}_R - \mathcal{E}_H^*, \omega\big)\Big[g^>\big(\mathcal{E}_H^*\big)g^<(\mathcal{E}_R) - g^<\big(\mathcal{E}_H^*\big)g^>(\mathcal{E}_R)\Big] \\ - h(\mathcal{E}_H + \mathcal{E}_R)\rho_L'(\mathcal{E}_H + \mathcal{E}_R, \omega)\Big[g^>(\mathcal{E}_H)g^>(\mathcal{E}_R) - g^<(\mathcal{E}_H)g^<(\mathcal{E}_R)\Big]\Big\}\Bigg].\end{aligned} \tag{6.69}$$

One can easily obtain the form of S_H by making the following substitutions in the above analysis $k_0 \rightarrow p_0$, $q_0 \rightarrow k_0$ and $k_0 \rightarrow p_0$. We can immediately infer a remarkable identity $S_L = S_H$. To find the expression for S_L we make the changes $k_0 \rightarrow q_0$, $q_0 \rightarrow p_0$, and $p_0 \rightarrow k_0$. This takes a little more care as we now have to make both contour integrals over the top half of the complex plane. This means we acquire an additional minus sign and it is easy to see that $S_L = -S_R$. These sources are the same rate multiplied by the same linear combination of chemical potentials. The fact that the rate is the same corresponds to the fact that all three sources are the same physical process. Later we will see that these chemical potentials are proportional to their number densities for a given approximation. The linear combination of chemical potentials then serves to couple the differential equations governing the number densities of the three particle species. The overall sign then serves to ensure that the sign of this source term is always negative, meaning that it relaxes the combination of chemical potentials, which is what we expect since it is CP conserving. Let us next turn these relaxation terms into single integrals. To begin with let us note that our spectral density functions ρ consist of four Dirac delta functions:

$$\delta(\omega_L + \omega_R - \omega_H),\ \delta(\omega_L + \omega_R + \omega_H),\ \delta(\omega_L - \omega_R - \omega_H),\ \delta(\omega_L - \omega_R + \omega_H). \tag{6.70}$$

The second delta function is never non-zero within the limits of integration and can be ignored. We can use the first and fourth delta functions to perform the integral over ω_H. Then insisting that the limits of integration are real we obtain step functions

$$\Theta(m_2 - m_3 - m_1) + \Theta(m_3 - m_2 - m_1). \tag{6.71}$$

Performing the same procedure with the third delta function gives a step function

$$\Theta(m_1 - m_2 - m_3) \tag{6.72}$$

to ensure a real value for $\omega_\pm$ in each case. Our new limits of integration are

$$\omega_\pm = \frac{|\Delta|\omega_R \pm k\sqrt{\Delta^2 + m_R^2 m_L^2}}{m_R^2}, \tag{6.73}$$

with $\Delta = \frac{1}{2}(M_L^2 + M_R^2 - M_H^2)$. We then can write the triscalar decay rate as

$$\begin{aligned}\Gamma_y = |\lambda|^2\beta \int_{m_R}^{\infty} \mathrm{d}\omega_R \int_{\omega^-}^{\omega^+} \mathrm{d}\omega \\ \{n_B(\omega_R)[1 + n_B(\omega_L)]n_B(\omega_L - \omega_R)[\Theta(m_R - m_L - m_H) - \Theta(m_L - m_R m_H)] \\ - n_B(\omega_R)n_B(\omega_L)[1 + n_B(\omega_L + \omega_R)]\Theta(m_H - m_R - m_L)\}.\end{aligned} \tag{6.74}$$

Finally performing the integral over $\mathrm{d}\omega$ we can reduce our integral to be over one variable

$$\begin{aligned}\Gamma_Y = -\frac{|\lambda|^2\beta}{16\pi^3} \int_{m_R}^{\infty} \mathrm{d}\omega_R h_B(\omega_R) \\ \times \left\{ \log\left(\frac{e^{\omega_R/T} - e^{\omega^+/T}}{e^{\omega_R/T} - e^{\omega^-/T}} \frac{e^{\omega^-/T} - 1}{e^{\omega^+/T} - 1} \right) \right. \\ [\Theta(m_1 - m_2 - m_\phi) - \Theta(m_2 - m_1 - m_\phi)] \\ \left. + \log\left(\frac{e^{-\omega_R/T} - e^{\omega^+/T}}{e^{-\omega_R/T} - e^{\omega^-/T}} \frac{e^{\omega^-/T} - 1}{e^{\omega^+/T} - 1} \right) \Theta(m_\phi - m_1 - m_2) \right\} \\ \equiv I_B(m_L, m_R, m_H)..\end{aligned} \tag{6.75}$$

Note that a triscalar decay cannot occur unless the parent species is larger than the combined mass of the daughter species giving these integrals a clear physical interpretation. Next let us show the analogous function for Yukawa interactions, skipping the lengthy and very similar derivation

$$\begin{aligned}\Gamma_Y = -\frac{|\lambda|^2\beta}{16\pi^3}\left(m_1^2 + m_2^2 - m_\phi^2\right) \int_{m_1}^{\infty} \mathrm{d}\omega_1 h_F(\omega_1) \\ \times \left\{ \log\left(\frac{e^{\omega_\phi^-/T} + e^{\omega_1/T}}{e^{\omega_\phi^+/T} + e^{\omega_1/T}} \frac{e^{\omega_\phi^+/T} - 1}{e^{\omega_\phi^-/T} - 1} \right) \right. \\ [\Theta(m_1 - m_2 - m_\phi) - \Theta(m_\phi - m_1 - m_2)] \\ \left. + \log\left(\frac{e^{\omega_\phi^-/T} + e^{-\omega_1/T}}{e^{\omega_\phi^+/T} - e^{-\omega_1/T}} \frac{e^{\omega_\phi^+/T} - 1}{e^{\omega_\phi^-/T} - 1} \right) \Theta(m_2 - m_1 - m_\phi) \right\} \\ \equiv |\lambda|^2 I_F(m_1, m_2, m_\phi)\end{aligned} \tag{6.76}$$

with $\omega_\phi^\pm$ given by

$$\omega_{\pm} = \frac{1}{2m_1^2}\left\{\omega_1\left|m_\phi^2 + m_1^2 - m_2^2\right| \right.$$
$$\left.\left[(\omega_1^2 - m_1^2)\left(m_1^2 - (m_2 + m_\phi)^2\right)\left(m_1^2 - (m_2 - m_\phi)^2\right]^{1/2}\right\}. \tag{6.77}$$

For a supergauge interaction such as

$$l \ni \phi\bar{\Psi}(g_L P_L + g_R P_R)\tilde{V}, \tag{6.78}$$

which is responsible for one-loop interactions such as the one shown in figure 6.6, the sources resulting from such interactions can be related to the fermion three-body rates [12]

$$S_{\tilde{V}} = (|g_L|^2 + |g_R|^2)I_F(m_\psi, m_\phi, m_{\tilde{V}}) + 2\,\mathrm{Re}\left(g_L g_R^*\right)\tilde{I}_F(m_\psi, m_\phi, m_{\tilde{V}}), \tag{6.79}$$

where we have defined

$$\tilde{I}_F(m_1, m_\phi, m_2) = \frac{2m_1 m_2}{m_1^2 + m_2^2 - m_\phi^2} I_F(m_1, m_\phi, m_2). \tag{6.80}$$

To conclude this section we will include all relevant one-loop three-body interactions in the MSSM, excluding the gauge–boson interactions. By relevant we mean all interactions that involve the order 1 coupling constants $\{y_t, g_1, g_2, g_3\}$. Since the three-body rates scale as the coupling constant squared, even three-body rates involving y_b and y_τ are heavily suppressed. In order to catalog the set of three-body rates it is useful to introduce a couple of shorthands. First let us denote

$$I_{F,B}^{1,2,3} = I_{F,B}(m_1, m_2, m_3), \tag{6.81}$$

as well as denoting the charged Higgs mixing angle as follows:

$$\begin{pmatrix} H_u^+ \\ H_d^- \end{pmatrix} = \begin{pmatrix} c_\alpha & s_\alpha \\ -s_\alpha & c_\alpha \end{pmatrix}\begin{pmatrix} H_1^+ \\ H_2^+ \end{pmatrix}. \tag{6.82}$$

We will also denote the left-handed quark doublet by Q_L for reasons we will explain in section 6.4. The set of three-body supergauge interactions between Higgs and Higgsinos are

$$\begin{aligned}\left\{\Gamma_{\tilde{V}}^{H_1,\tilde{H}}, \Gamma_{\tilde{V}}^{H_2\tilde{H}}\right\} = \Bigg\{ & g_1^2\left(I_F^{\tilde{H},H_1,\tilde{B}} - s_\alpha \tilde{I}_F^{\tilde{H},H_1,\tilde{B}}\right) \\ & + 3g_2^2\left(I_F^{\tilde{H},H_1,\tilde{W}} + s_\alpha \tilde{I}_F^{\tilde{H},H_1,\tilde{W}}\right), g_1^2\left(I_F^{\tilde{H},H_2,\tilde{B}} + s_\alpha \tilde{I}_F^{\tilde{H},H_2,\tilde{B}}\right) \\ & + 3g_2^2\left(I_F^{\tilde{H},H_2,\tilde{W}} - s_\alpha \tilde{I}_F^{\tilde{H},H_2,\tilde{W}}\right)\Bigg\}.\end{aligned} \tag{6.83}$$

The set of three-body supergauge interactions involving third generation quarks are

$$\begin{aligned}&\left\{\Gamma_{\tilde{V}}^{Q_L,\tilde{Q}_L},\ \Gamma_{\tilde{V}}^{t_L,\tilde{t}_R}\right\} = N_C \times \\ &\left\{\frac{g_1^2}{9} I_F^{Q_L,\tilde{Q}_L,\tilde{B}} + 3g_2^2 I_F^{Q_L,\tilde{Q}_L,\tilde{W}}\right. \\ &\left. + \frac{2(N_C^2-1)}{N_C} g_3^2 I_F^{Q_L,\tilde{Q}_L,\tilde{G}},\ \frac{8g_1^2}{9} I_F^{t_R,\tilde{t}_R,\tilde{B}} + \frac{(N_c^2-1)}{N_C} g_3^2 I_F^{t_R,\tilde{t}_R,\tilde{G}}\right\},\end{aligned} \tag{6.84}$$

and finally the three-body rates involving third-generation interactions between squarks, quarks, as well as Higgs and Higgsinos are

$$\begin{aligned}&\left\{\Gamma_Y^{\tilde{t}_R,\tilde{Q}_L,H_1},\ \Gamma_Y^{\tilde{t}_R,\tilde{Q}_L,H_2},\ \Gamma_Y^{\tilde{t}_R,Q_L,\tilde{H}},\ \Gamma_Y^{t_R,Q_L,H_1},\ \Gamma_Y^{t_R,Q_L,H_2},\ \Gamma_Y^{t_R,\tilde{Q}_L,\tilde{H}}\right\} = 2y_t^2 N_C \\ &\left\{|s_\alpha \mu^* + c_\alpha A_t|^2 I_B^{\tilde{t}_R,\tilde{Q}_L,H_1},\ |c_\alpha \mu^* - s_\alpha A_t|^2 I_B^{\tilde{t}_R,\tilde{Q}_L,H_2},\ I_F^{\tilde{H},\tilde{t}_R,Q_L},\ c_\alpha^2 I_F^{t_R,Q_L,H_1},\right. \\ &\left. s_\alpha^2 I_F^{t_R,Q_L,H_2},\ I_F^{t_R,\tilde{H},\tilde{Q}_L}\right\}.\end{aligned} \tag{6.85}$$

6.2.5 Four-body interactions

Four-body scattering processes such as the one depicted in figure 6.5 are typically small compared to the three-body rates calculated in the previous subsection. However, there are regions of parameter space where a three-body rate is kinematically forbidden and the four-body rate dominates. Indeed if all relevant three-body rates are kinematically suppressed, then ignoring four-body scattering processes would result in particle number densities not relaxing in the symmetric phase. An estimate of the four-body rate for the four-body scattering process involving right-handed tops, Higgs, and a third-generation left-handed quark doublet is [13]

$$\Gamma_{\text{scatt}}^{t,\,H,\,Q} = 0.129\frac{g_3^2}{4\pi}T^2. \tag{6.86}$$

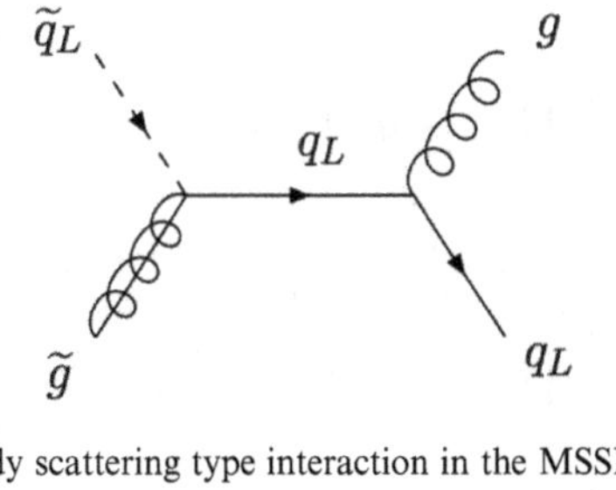

Figure 6.5. Example of a four-body scattering type interaction in the MSSM. When the three-body rates are kinematically forbidden or heavily suppressed the four-body scattering rates dominate.

6.3 Local equilibrium approximations

If an interaction is very fast compared to an appropriate time scale (the diffusion time) then one can simplify one's set of transport equations by setting the linear combination of chemical potentials involved in the interaction to zero. Also, if all decay rates from a strongly sourced particular particle species are very slow, you can decouple their transport equations. These two types of steps are the essence of most of the approximations made to make the problem of describing the evolution of particle number densities during the electroweak phase transitions. We will review the most commonly used ones here.

6.4 Gauge and supergauge equilibrium

Gauge interactions are typically assumed to be fast enough to equilibrate gauge multiplets. For example consider an $SU(2)_L$ doublet denoted by $(\phi_\uparrow, \phi_\downarrow)^T$. If the rate of this interaction with W bosons is sufficiently fast, then the W boson chemical potential is equal to the $SU(2)_L$ triplet

$$\mu_W = \mu_{\phi_\uparrow} - \mu_{\phi_\downarrow}. \tag{6.87}$$

Moreover every $SU(2)_L$ triplet is equal to each other since they are equal to the W boson. Usually the W boson is assumed to have null chemical potential since we are mostly concerned with the dynamics of the symmetric phase near the bubble wall. This is because electroweak sphalerons, which are responsible for producing the baryon asymmetry, are only active in the symmetric phase and the chemical potentials of all $SU(2)_L$ gauge bosons are zero by symmetry in that phase. This results in all gauge multiplet chemical potentials being equal to zero. Furthermore, gauge triplets can be decoupled from the transport equations as follows. Consider a supergauge interaction, such as the ones depicted in figure 6.6. The source with the left-handed (s)bottom and (s)top is given by

$$\begin{aligned} \partial_\mu j^\mu_{t_L} &\sim g_2^2 |V_{tb}|^2 I_F^{t_L, \tilde{b}_L, \tilde{W}^\pm}(\mu_{t_L} - \mu_{\tilde{b}_L} - \mu_{\tilde{W}^\pm}) \\ \partial_\mu j^\mu_{b_L} &\sim g_2^2 |V_{tb}|^2 I_F^{b_L, \tilde{t}_L, \tilde{W}^\pm}(\mu_{b_L} - \mu_{\tilde{t}_L} - \mu_{\tilde{W}^\pm}). \end{aligned} \tag{6.88}$$

The $SU(2)_L$ singlet and triplet transport equations are then

$$\begin{aligned} \partial_\mu\left(j_{t_L} \pm j_{b_L}\right)^\mu = &-\frac{N_C g_2^2}{2}|V_{tb}|^2\left[\Gamma^{t_L,\tilde{b}_L,\tilde{W}} \pm \Gamma^{b_L,\tilde{t}_L,\tilde{W}}\right] \\ &\left[\left(\mu_{t_L} + \mu_{b_L}\right) - \left(\mu_{\tilde{t}_L} + \mu_{\tilde{b}_L}\right)\right] \\ &- \frac{N_C g_2^2}{2}|V_{tb}|^2\left[\Gamma^{t_L,\tilde{b}_L,\tilde{W}} \mp \Gamma^{b_L,\tilde{t}_L,\tilde{W}}\right] \\ &\left[\left(\mu_{t_L} - \mu_{b_L}\right) - \left(\mu_{\tilde{t}_L} - \mu_{\tilde{b}_L}\right) - 2\mu_{\tilde{W}}\right]. \end{aligned} \tag{6.89}$$

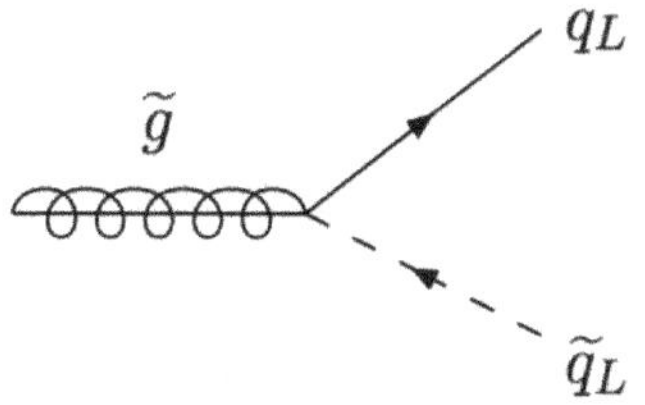

Figure 6.6. Example of a supergauge interaction in the MSSM, or indeed any supersymmetric extension of the Standard Model.

Note that if $\Gamma^{t_L,\,\tilde{b}_L,\,\tilde{W}} \approx \Gamma^{b_L,\,\tilde{t}_L,\,\tilde{W}}$ the gauge single transport equation decouples from the dynamics of the gauge triplet. This suppresses the effects of gauge triplets further to the arguments made before and we can, in general, just consider the network of transport equations for gauge singlets. Finally if supergauge rates such as the ones depicted in figure 6.6 are sufficiently fast, then particles are in equilibrium with their superpartners and we can treat them as one species. Even if they are not, the set of three-body interactions suppress deviations from supergauge equilibrium. To see this consider the case where all three-body rates are in equilibrium. In such a case if we treat left-handed doublets as a single species (denoted by $\tilde{Q}$ and Q for the (s) quark doublet) we have

$$\begin{aligned} &\mu_Q + \mu_{H_1} - \mu_t = \mu_Q + \mu_{H_2} - \mu_t = \mu_{\tilde{Q}} + \mu_{H_1} - \mu_{\tilde{t}} \\ &= \mu_{\tilde{Q}} + \mu_{H_2} - \mu_{\tilde{t}} = \mu_Q + \mu_{\tilde{H}} - \mu_{\tilde{t}} = \mu_{\tilde{Q}} + \mu_{\tilde{H}} - \mu_t \\ &= 0. \end{aligned} \tag{6.90}$$

Simple manipulation of the above equations then implies supergauge equilibrium even if the supergauge rates are not particularly fast,

$$\mu_Q = \mu_{\tilde{Q}} \quad \mu_t = \mu_{\tilde{t}} \quad \mu_{H_1} = \mu_{H_2} = \mu_{\tilde{H}}. \tag{6.91}$$

So typically both Higgs doublets and the Higgsino are treated as a single-particle species and gauge singlets and their superpartners are treated as a single-particle species.

6.5 Fast rate approximations

The first two generations of baryons are very weakly sourced as their Yukawa couplings with the Higgs are small. This means there is an approximate baryon number conservation (up to weak sphaleron interactions) for the first two generations. The consequence of this is we can write

$$\mu_{Q_iL} = -2\mu_{d_iR} = -2\mu_{u_iR} \; \{i \in 1, 2\}. \tag{6.92}$$

By contrast top Yukawa couplings are large, so it is common to approximate that the one-loop Yukawa rate equilibrates the linear combination of chemical potentials

$$\mu_Q = \mu_T - \mu_H. \tag{6.93}$$

Also the strong sphaleron rate is very large, about a thousand times larger than the weak sphaleron rate. This has inspired people to make the approximation that

$$\mu_5 = \mu_L - \mu_R = 2\mu_{Q_iL} - \mu_{d_iR} - \mu_{u_iR} \; \{i \in (1, 2, 3)\} \approx 2\mu_Q - \mu_T + 9\mu_b = 0, \quad (6.94)$$

where we have used local baryon number conservation for the first two generations. These two approximations can greatly simplify our set of transport equations. Unfortunately, they can give an estimate of the baryon asymmetry that can differ by up to two orders of magnitude from a more precise calculation [11], so we present and later use these approximations for pedagogical purposes only before giving a detailed analytic approach that does not use these approximations.

References

[1] Babu K S, Mohapatra R N and Nasri S 2006 *Postsphaleron Baryogenesis* **97** 131301

[2] Riotto A 1998 More relaxed supersymmetric electroweak baryogenesis *Phys. Rev.* D **58** 095009

[3] Huet P and Nelson A E 1996 Electroweak baryogenesis in supersymmetric models *Phys. Rev.* D **53** 8

[4] Carena M, Quiros M, Seco M and Wagner C E M 2003 Improved results in supersymmetric electroweak baryogenesis *Nucl. Phys.* B **650** 1

[5] Levinson E J and Boal D H 1985 Self-energy corrections to fermions in the presence of a thermal background *Phys. Rev.* D **31** 3280

[6] Kalashnikov O K 1999 Fermi spectra and their gauge invariance in hot and dense abelian and non-abelian theories *Phys. Scri.* **60** 131

[7] Lee C, Cirigliano V and Ramsey-Musolf M J 2005 Resonant relaxation in electroweak baryogenesis *Phys. Rev.* D **71** 7

[8] Cirigliano V *et al* 2010 Flavored quantum Boltzmann equations *Phys. Rev.* D **81** 10

[9] Weldon H A 1989 Dynamical holes in the quark-gluon plasma *Phys. Rev.* D **40** 2410

[10] Weldon H A 2000 Structure of the quark propagator at high temperatures *Phys. Rev.* D **61** 036003

[11] Cirigliano V, Ramsey-Musolf M J, Tulin S and Lee C 2006 Yukawa and triscalar processes in electroweak baryogenesis *Phys. Rev.* D **73** 115009

[12] Chung D J H, Garbrecht B, Ramsey-Musolf M J and Tulin S 2009 Supergauge interactions and electroweak baryogenesis *JHEP* JHEP12(2009)067

[13] Joyce M, Prokopec T and Turok N 1996 Nonlocal electroweak baryogenesis. I. thin wall regime *Phys. Rev.* D **53** 2930

Graham Albert White

Chapter 7

Plasma and bubble dynamics

There are a number of nuisance variables that we have left undefined up to this point that enter into solving the baryon asymmetry:

- the bubble wall velocity, v_w;
- the diffusion coefficients for each particle, D_i;
- the thermal widths, Γ_i; and
- the thermal masses ΔM_{Ti}.

It is the evaluation of these parameters that we cover in this chapter. Many of these calculations are more conveniently calculated in the finite temperature imaginary time formalism, which is appropriate for systems in equilibrium. Corrections to these parameters due to the departure from equilibrium is assumed to be insignificant. We therefore begin this section with a brief summary of the imaginary time formalism.

7.1 Imaginary time formalism

Fortunately our treatment of the imaginary time formalism can be fairly shallow if our goal is just to give the minimum background necessary to make calculations in this section self-contained. We will therefore merely derive momentum space propagators for both bosons and fermions. Recall that for a system in thermal equilibrium the density matrix is just $Z^{-1}e^{-\beta\mathcal{H}}$. The expectation value of any operator $\mathcal{O}$ is then

$$\langle\mathcal{O}\rangle = Z^{-1}\,\mathrm{Tr}\,[\rho\mathcal{O}]. \tag{7.1}$$

doi:10.1088/978-1-6817-4457-5ch7

Consider a two point correlator [1]

$$\begin{aligned}\langle\phi(t)\phi(t')\rangle &= Z^{-1}\,\mathrm{Tr}\left[\mathrm{e}^{-\beta H}\phi(t)\phi(t')\right]\\ &= Z^{-1}\,\mathrm{Tr}\left[\mathrm{e}^{-\beta H}\phi(t)\mathrm{e}^{\beta H}\mathrm{e}^{-\beta H}\phi(t')\right]\\ &= Z^{-1}\,\mathrm{Tr}\left[\phi(t+\mathrm{i}\beta)\mathrm{e}^{-\beta H}\phi(t')\right]\\ &= \langle\phi(t')\phi(t+\mathrm{i}\beta)\rangle.\end{aligned} \tag{7.2}$$

This identity is known as the Kubo–Martin–Schwinger (KMS) relation [2, 3]. It is a statement that there is an inherent periodicity in the Green's functions at finite temperature that arises from the fact that the density matrix is the time evolution operator for imaginary time β. Note that the above equations implied that the derivation was for bosons but no step assumed this and it applies equally to fermions. Let us Wick rotate to imaginary time. The two-point Green's functions therefore obey the periodicity relation for $\tau > 0$

$$G_\beta(0,\tau) = \pm\, G_\beta(\beta,\tau). \tag{7.3}$$

A Green's function that is periodic in a space or time variable has a Fourier transform that is a sum over a discrete set of frequencies. We can therefore write for frequencies $\omega_n = (n + n_{\mathrm{F/B}})\pi/\beta$, where $n_{\mathrm{F/B}} = (1, 0)$ for fermions and bosons, respectively, [4]

$$G_\beta(\tau) = \frac{1}{\beta}\sum_n \mathrm{e}^{-\mathrm{i}\omega\tau} G_\beta(\omega_n) \tag{7.4}$$

$$G_\beta(\omega_n) = \frac{1}{2}\int_{-\beta}^{\beta} \mathrm{d}\tau \mathrm{e}^{\mathrm{i}\omega\tau} G_\beta(\tau). \tag{7.5}$$

These frequencies are known as the Matsubara frequencies. The fact that imaginary time becomes periodic at finite temperature is a manifestation of the fact that the temperature that defines the field is not Lorentz invariant. The momentum space Green's functions for bosons and fermions are then

$$\begin{aligned}G_\beta\left(\vec{k},\omega_n\right) &= \frac{1}{\omega_n^2 + \vec{k} + m^2}\\ S_\beta\left(\vec{k},\omega_n\right) &= \frac{\gamma^0\omega_n + \vec{\gamma}\cdot\vec{k} - m}{\omega_n^2 + k^2 + m^2},\end{aligned} \tag{7.6}$$

respectively.

7.2 Diffusion coefficients

Consider a particle in a system that is out of equilibrium whose number density can be denoted by the function $n(x, t)$. Suppose at a given time $t + \mathrm{d}t$ a particle is at position x, then at a time t the particle must be at the position $x - \mathrm{d}x$. In other words [5]

$$
\begin{aligned}
n(t + \mathrm{d}t, x) &= \langle n(t, x - \mathrm{d}x) \rangle \\
&= n(t, x) - \frac{\partial n}{\partial x} \langle \delta x \rangle + \frac{1}{2} \frac{\partial^2 n}{\partial x^2} \langle \delta x^2 \rangle .
\end{aligned}
\tag{7.7}
$$

Let us assume that our system is isotropic so $\langle \delta x \rangle = 0$. The above equation reduces to

$$
\frac{\partial n}{\partial t} = D \frac{\partial^2 n}{\partial x^2}
\tag{7.8}
$$

for some diffusion coefficient D. The particle number current $J_\mu = (n, -D\vec{\nabla} n)$ implicitly defines the diffusion coefficient via the proportionality of the three-current $\vec{J}$ being proportional to the gradient of the number density[1]

$$
\vec{J} = -D\vec{\nabla} n .
\tag{7.9}
$$

The three-current density can just be defined in the usual way using the distribution function, f, to define the average of the velocity

$$
\begin{aligned}
\vec{J} = \int \mathrm{d}^3 p f(p) \vec{v} &= \int \mathrm{d}^3 p (f_0 + \delta f) \vec{v} \\
&= \int \mathrm{d}^3 p \delta f \vec{v} \\
&= \int \mathrm{d}^3 p \delta f \frac{\vec{p}}{E} .
\end{aligned}
\tag{7.10}
$$

Here $f_0 = (\exp[\beta(p_0 - \mu)] \pm 1)^{-1}$ are the equilibrium distribution functions for chemical potential μ and fermions (+) and bosons (−). Let us calculate the diffusion coefficients for fermions only. We can relate the number density to the chemical potentials as follows (see section 9 for details):

$$
n \approx n_0 - \mu \frac{T^2}{12} ,
\tag{7.11}
$$

with n_0 the equilibrium number density. So to find the diffusion coefficient we ultimately need to find a way of solving for δf as well as the chemical potentials. Let us make use of classical Boltzmann equations, in this case the static one which sets $\frac{\partial f}{\partial t} = 0$

$$
\frac{\mathrm{d}f}{\mathrm{d}t} = \frac{\vec{p}}{p_0} \cdot \vec{\nabla} f = -C(f) .
\tag{7.12}
$$

We assume that $\vec{\nabla} f \to \frac{\partial f}{\partial x}$ and rewrite the left-hand side of the above equation as [7, 8–11]

[1] For a nice pedagogical introduction to these concepts see [6].

$$T\frac{p_x}{p_0}\frac{\partial\mu}{\partial x}\frac{\partial f}{\partial\left(\beta\left[p_0 - E\right]\right)}. \tag{7.13}$$

This gives us a potential avenue to relate the chemical potential (or rather its derivative) to other calculable quantities. Let us initially concentrate on calculating the diffusion coefficients of gauge bosons. The collision integral is defined as a sum of four-legged fermionic scattering diagrams with gauge bosons as intermediate states

$$\begin{aligned} C(f) = \int \frac{\mathrm{d}^3k\mathrm{d}^3p'\mathrm{d}^3k'}{(2\pi)^9 8E_kE_{p'}E_{k'}}|M(s, t, u)|^2(2\pi)^2\delta(p + k - p' - k') \\ \times\left[f_pf_k\left(1 - f_{p'}\right)\left(1f_{k'}\right) - f_{p'}f_{k'}\left(1 - f_p\right)\left(1 - f_k\right)\right]. \end{aligned} \tag{7.14}$$

Here $|M|^2$ is the appropriate modulus squared scattering amplitude and (s, t, u) are the usual Mandelstam variables. Let us ignore the fluctuations of the second fluid setting $\delta f_k = \delta f_{k'} = 0$. Let us also assume that little energy is transferred in a collision so $k \sim k'$ and $p \sim p'$. In this case the collision integral simplifies to

$$\begin{aligned} C(f) = \int \frac{\mathrm{d}^3k\mathrm{d}^3p'\mathrm{d}^3k'}{(2\pi)^9 8E_kE_{p'}E_{k'}}|M(s, t, u)|^2(2\pi)^2\delta(p + k - p' - k') \\ \times\left(\delta f_p - \delta f_{p'}\right)f_k^0\left(1 - f_k^0\right). \end{aligned} \tag{7.15}$$

To solve the static Boltzmann equation we assume the solution is of the form

$$\delta f_p = g(p, x)\frac{p_x}{p_0}. \tag{7.16}$$

Then, using the approximation that energy transfer is low we can write

$$\delta f_p - \delta f_{p'} \approx g(p, x)\frac{\left(p^2 - p\cdot p'\right)}{p^2}. \tag{7.17}$$

Our Boltzmann equation then the takes the form

$$\begin{aligned} T\frac{\partial\mu}{\partial x}\frac{\partial f}{\partial\beta\left[p_0 - \mu\right]} &= \frac{-1}{p_0^3}\int \frac{\mathrm{d}^3k\mathrm{d}^3p'\mathrm{d}^3k'}{(2\pi)^9 8E_kE_{p'}E_{k'}}|M(s, t, u)|^2(2\pi)^2\delta(p + k - p' - k') \\ &\quad\times g(p, x)p\cdot p'\left[\frac{-\mathrm{e}^{\beta k_0}}{(\mathrm{e}^{\beta k_0} + 1)^2}\right] \\ &\equiv -g(p, x)\Gamma_t, \end{aligned} \tag{7.18}$$

where $g(p, x)$ has been taken out of the integral over (p', k, k'). Next substitute our ansatz into our definition of the three current density. Using the relations that $\vec{\nabla}n \to \frac{\partial n}{\partial x} \propto \frac{\partial\mu}{\partial x}$ we can eliminate the function $g(p, x)$ by combining the definition of the three-current with our Boltzmann equations simplifying the calculation dramatically

$$\int d^3p\delta f\frac{p}{E} = -D\nabla\vec{n}$$

$$\int d^3g(p, x)\frac{p_x^2}{E^2} = D\frac{T^2}{12}\frac{\partial\mu}{\partial x}$$

$$D = \frac{12}{T^2}\int d^3pg(p, x)\frac{p_x^2}{E^2}\left(\frac{\partial\mu}{\partial x}\right)^{-1} \tag{7.19}$$

$$= \frac{12}{T^3}\int d^3p\left(\frac{p_x^2}{E^2}\right)\frac{1}{\Gamma_t}\frac{\partial f}{\partial\beta[E-\mu]}.$$

To work out an example consider the case where we have Standard Model gauge interactions. Assuming $p^2 = k^2 = \cdots = 0$, the relevant modulus squared scattering amplitudes are

$$|M_W|^2 = \frac{144 g_2^4(p\cdot k)^2}{\left(2p\cdot p' + M_w^2\right)^2}$$

$$|M_B|^2 = \frac{320 Y^2 g_1^4(p\cdot k)^2}{\left(2p\cdot p' + M_B^2\right)^2} \tag{7.20}$$

$$|M_G|^2 = \frac{128 g_3^4(p\cdot(k+k'))^2}{\left(2p\cdot p' + M_B^2\right)^2},$$

with thermal masses

$$M_B^2 = \frac{g_2^2\tan^2\theta_W T^2}{3} \quad M_W^2 = \frac{5g_2^2T^2}{3} \quad M_G^2 = 2g_3^2T^2. \tag{7.21}$$

Let us calculate Γ_t in the center-of-mass frame and calculate only to first-order in M/T. The gluon case gives, for example,

$$\Gamma_t \approx \frac{1}{6\pi}\frac{T^3}{p^2}\log\left(\frac{4pTx_2}{M^2}\right). \tag{7.22}$$

We then have a pretty nasty integral to perform that includes a logarithm in the integrand. We can approximate this integral by taking the logarithm out of the integrand and evaluating it at the point p which maximizes $p^6\frac{\partial f}{\partial\beta[E-\mu]}$. This gives

$$
\begin{aligned}
D_B^{-1} &\approx \frac{100}{7\pi}\alpha_W^2 \tan^4\theta_W Y^2 T \log\left(\frac{32T^2}{M_B^2}\right) \\
D_W^{-1} &\approx \frac{45}{7\pi}\alpha_W^2 T \log\left(\frac{32T^2}{M_W^2}\right) \\
D_G^{-1} &\approx \frac{80}{7\pi}\alpha_S^2 T \log\left(\frac{32T^2}{M_G^2}\right).
\end{aligned}
\tag{7.23}
$$

The diffusion coefficients of fermions in the Standard Model is just then the sum of the contributions from respective gauge bosons

$$
\begin{aligned}
D_{l_R}^{-1} &= D_B^{-1} \approx \frac{T}{380} \\
D_{l_L}^{-1} &= D_W^{-1} + \frac{1}{4}D_{l_R}^{-1} \approx \frac{T}{100} \\
D_{q_{L,R}}^{-1} &= Y^2 D_{l_R}^{-1} + \epsilon_{L,R} D_W^{-1} + D_G^{-1} \approx \frac{T}{6},
\end{aligned}
\tag{7.24}
$$

where $\epsilon_{L,R} = (1, 0)$.

7.3 Thermal widths

The poles of a propagator are shifted in zero temperature field theory by an amount equal to the expectation value of imaginary part of the self-energy

$$
\Gamma = \langle \mathrm{Im}(\Sigma(P)) \rangle. \tag{7.25}
$$

At finite temperature we do the same using finite temperature propagators. The self-energy is defined along the real axis, $\Sigma(P)$. Let $f(P)^{-1} = \Sigma(P)$, the imaginary part of the self-energy is given by [12]

$$
\mathrm{Im}\,[\Sigma(P)] = \frac{1}{2i}\left(\frac{1}{f(p + \mathrm{i}\epsilon)} - \frac{1}{f(p - \mathrm{i}\epsilon)}\right). \tag{7.26}
$$

There are two types of contribution to this, residues of poles and discontinuities across the real axis [13–15]. In practice the thermal width is dominated by gauge interactions. For the MSSM the thermal mass has been found in various limits. In the limit where fermions have masses that are heavy compared to the temperature the contribution is [16]

$$
\Gamma = \frac{g^2 T C_f}{8\pi}, \tag{7.27}
$$

where C_f is the Casimir invariant of the fermion representation. In the massless fermion limit at rest, the contribution is the same as the above with the coefficient $2a(N, N_F)$ which depends on the gauge group and the number of fermions. Finally the $p > 0$ massless fermion limit gives a small velocity-dependent contribution that

Table 7.1. Thermal widths relevant to the MSSM. Values taken from [16].

Relevant thermal widths in the MSSM	
Particle	Thermal width
$\Gamma_{\widetilde{H}}$	$0.025T$
$\Gamma_{\widetilde{B}}$	$0.003T$
$\Gamma_{\widetilde{W}}$	$0.065T$
Γ_t	$0.5T$
$\Gamma_{\tilde{t}}$	$0.5T$
Γ_t	$0.16T$
$\Gamma_{\tilde{t}}$	$0.16T$
Γ_τ	$0.002T$
$\Gamma_{\tilde{\tau}}$	$0.002T$

in general can be ignored. The thermal width of fermions interacting with scalars is generally quite small. For the Higgsino–stop interaction the contribution is $\sim 0.01Ty^2$ in the massless limit and $\lesssim 0.0025Ty^2$ for $M = O(T/2)$ for scalar mass M and Yukawa coupling y. The thermal width of fermions with gauge symmetry is therefore $\approx \frac{g^2 TC_f}{4\pi}$, where the variation in $a(N, N_F)$ and the contribution from moving massless fermions is ignored. A table of values of all thermal widths relevant to baryon asymmetry calculation in the MSSM is given in table 7.1.

7.4 Thermal masses

We now turn our attention to calculating the one loop thermal correction to the mass, known as the Debye mass [1, 17]. Let us start off by calculating the easiest example. Consider the model

$$\mathcal{L} = \frac{1}{2}\partial_\mu\phi\partial^\mu\phi - \frac{m^2}{2}\phi^2 - \frac{\lambda}{4!}\phi^4. \tag{7.28}$$

In the imaginary time formalism the integral over k_0 is replaced with a sum over Matsubara frequencies. The diagram in figure 7.1 is

$$\Delta m^2 = \frac{\lambda}{2\beta}\sum_n \int \frac{d^3k}{(2\pi)^3}\frac{1}{\left(\frac{2n\pi}{\beta}\right)^2 + k^2 + m^2}. \tag{7.29}$$

Note that the sum $(n^2 + y^2)^{-1} = (\pi/y)\coth \pi y$, which allows us to write the above as

$$\Delta m^2 = \frac{\lambda}{4}\int \frac{d^3k}{(2\pi)^3}\frac{1}{\sqrt{k^2 + m^2}}\coth\frac{\beta}{2}\sqrt{k^2 + m^2}. \tag{7.30}$$

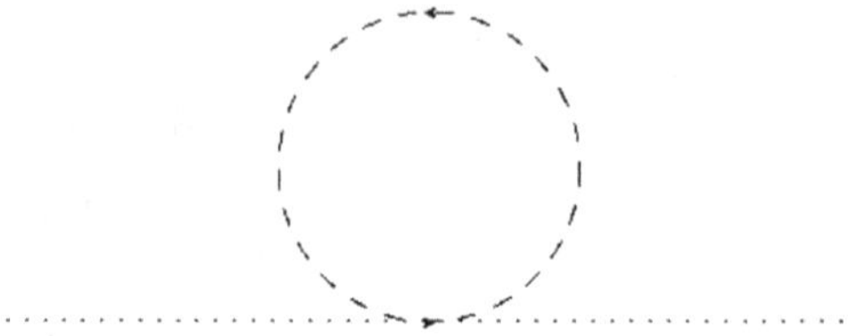

Figure 7.1. Relevant diagram for calculating the thermal mass correction for a scalar field theory with ϕ^4 interactions.

We can separate this into $0T$ and thermal corrections using the definition of the hyperbolic trigonometric function $\coth \beta y = 1 + 2n_B(2y)$ and write

$$\Delta m^2 = \frac{\lambda}{4}\int \frac{d^3k}{(2\pi)^3}\frac{1}{\sqrt{k^2+m^2}} + \frac{\lambda}{2}\int \frac{d^3k}{(2\pi)^3}\frac{1}{\sqrt{k^2+m^2}}\frac{1}{e^{\beta\sqrt{k^2+m^2}}-1}. \tag{7.31}$$

Let us denote the second term by Δm_T^2. In the limit where $m \to 0$ (or if the mass is small compared to the temperature) then we can make the approximation

$$\begin{aligned}\Delta m_T^2 = \frac{\lambda}{2}\int \frac{d^3k}{(2\pi)^3}\frac{1}{\sqrt{k^2+m^2}}\frac{1}{e^{\beta\sqrt{k^2+m^2}}-1} &\approx \frac{4\pi\lambda}{2}\int_0^\infty \frac{p^2 dp}{(2\pi)^3}\frac{1}{p}\frac{1}{e^{\beta p}-1} \\ &= \frac{4\pi\lambda}{2}\int \frac{dp}{(2\pi)^3} p e^{\beta p} - 1 \\ &= \frac{\lambda}{4\pi^2\beta^2}\int_0^\infty \frac{p}{e^p - 1} \\ &= \frac{\lambda T^2}{24},\end{aligned} \tag{7.32}$$

where in the last line we have used the identity $\int p/(e^p - 1) = \pi^2/6$. In table 7.2 the thermal masses for the MSSM are given, separated into their standard model and BSM contributions.

7.5 Bubble wall velocity

Calculating the bubble wall velocity is a difficult task and there are essentially two approaches that tend to be used. One approach requires the use of multiple approximations including at one stage treating the particles in the plasma classically. This approach is not always viable, for example if multiple fields have a space–time varying vacuum during a phase transition the approximations we will introduce in this section are not necessarily valid and one needs to follow a purely numerical approach [19–21]. The conventional semi-analytic approach begins by deriving the classical equations of motion [22, 23]. For a single scalar field this is

Table 7.2. Relevant thermal masses for electroweak baryogenesis (EWBG) in the MSSM separated into Standard Model contributions (A), interactions with Higgsinos, binos, winos, and RH stops (B) and RH sbottoms and staus (C). Table reproduced from [18].

Thermal masses in the MSSM			
Particle	$\delta m_{\mathrm{SM}}/T^2$ (A)	$\delta m_{\mathrm{SUSY}}/T^2$ (B)	$\delta m_{\mathrm{SUSY}}/T^2$ (C)
q_L	$\frac{1}{6}g_3^2 + \frac{3}{32}g_2^2 + \frac{1}{288}g_1^2 + \frac{1}{16}y_t^2 + \frac{1}{16}y_b^2$	$+\frac{1}{16}y_t^2$	$+\frac{1}{16}y_b^2$
t_R	$\frac{1}{6}g_3^2 + \frac{1}{18}g_1^2 + \frac{1}{8}y_t^2$	$+\frac{1}{18}g_1^2$	
b_R	$\frac{1}{6}g_3^2 + \frac{1}{72}g_1^2 + \frac{1}{8}y_b^2$		$+\frac{1}{72}g_1^2$
l_L	$\frac{3}{32}g_2^2 + \frac{1}{32}g_1^2 + \frac{1}{16}y_\tau^2$		$\frac{1}{16}y_\tau^2$
τ_R	$\frac{1}{8}g_1^2 + \frac{1}{8}y_\tau^2$		$+\frac{1}{8}g_1^2$
$\tilde{H}_u$		$\frac{3}{16}g_2^2 + \frac{1}{16}g_1^2 + \frac{3}{16}y_t^2$	
$\tilde{H}_d$		$\frac{3}{16}g_2^2 + \frac{1}{16}g_1^2$	$+\frac{3}{16}y_b^2 + \frac{1}{16}y_\tau^2$
$\tilde{W}$		$\frac{3}{8}g_2^2$	
$\tilde{B}$		$\frac{5}{12}g_2^2$	$+\frac{1}{6}g_2^2$
$\tilde{t}_R$		$\frac{4}{9}g_3^2 + \frac{1}{3}g_1^2 + \frac{1}{3}y_t^2$	$-\frac{1}{9}g_1^2$
$\tilde{b}_R$			$\frac{4}{9}g_3^2 + \frac{1}{18}g_1^2 + \frac{1}{3}y_b^2$
$\tilde{\tau}_R$			$\frac{1}{2}g_1^2 + \frac{1}{3}y_\tau^2$
H_u	$\frac{3}{16}g_2^2 + \frac{1}{16}g_1^2 + \frac{1}{4}y_t^2$	$\frac{3}{16}g_2^2 - \frac{1}{48}g_1^2 + \frac{1}{4}y_t^2$	$\frac{1}{12}g_1^2$
H_d	$\frac{3}{16}g_2^2 + \frac{1}{16}g_1^2 + \frac{1}{4}y_t^2 + \frac{1}{12}y_\tau^2$	$\frac{3}{16}g_2^2 + \frac{7}{48}g_1^2$	$-\frac{1}{12}g_1^2 + \frac{1}{4}y_b^2 + \frac{1}{12}y_\tau^2$

$$\Box\phi - \frac{\partial V}{\partial \phi} - \sum \frac{\mathrm{d}m_A^2}{\mathrm{d}\phi}A^2 + \frac{\mathrm{d}m_\Psi}{\mathrm{d}\phi}\overline{\Psi}_R\Psi_L, \tag{7.33}$$

where we have suppressed gauge indices. We set the scalar field to a classical value with symmetric perturbations. That is $\phi \to \phi_{\mathrm{cl}} + \delta\phi$ with $\langle \delta\phi \rangle = 0$. We then take the trace of the finite temperature density matrix to obtain

$$\Box\phi_{\mathrm{cl}} - \frac{\partial V}{\partial \phi}\bigg|_{\mathrm{cl}} - \frac{\mathrm{d}m_\phi^2}{\mathrm{d}\phi}\langle \delta\phi^2 \rangle - \sum \frac{\mathrm{d}m_A^2}{\mathrm{d}\phi}\langle A^2 \rangle + \frac{\mathrm{d}m_\Psi}{\mathrm{d}\phi}\left\langle \overline{\Psi}_R\Psi_L \right\rangle. \tag{7.34}$$

If one assumes WKB waves then one can write the second moment of each field as

$$\begin{aligned} \langle A^2 \rangle &= \langle A^2 \rangle_{\mathrm{vac}} + \sum \int \frac{\mathrm{d}^3k}{(2\pi)^3 E} f(k, x) \\ \left\langle \overline{\Psi}_R\Psi_L \right\rangle &= \frac{1}{2}\left\langle \overline{\Psi}_R\Psi_L \right\rangle_{\mathrm{vac}} + \sum \int \frac{\mathrm{d}^3k}{(2\pi)^3}\frac{m}{2E} f(k, x), \end{aligned} \tag{7.35}$$

where we have omitted the group and Dirac indices. Using the trivial identity that $m\frac{\mathrm{d}m}{\mathrm{d}\phi} = \frac{1}{2}\frac{\mathrm{d}m^2}{\mathrm{d}\phi}$ and omitting the subscript 'cl', we can write the equations of motion in the form

$$\Box\phi - \frac{\partial V}{\partial\phi} + \sum\frac{\mathrm{d}m^2}{\mathrm{d}\phi}\int\frac{\mathrm{d}^3k}{(2\pi)^3 2E}f(k, x) = 0. \tag{7.36}$$

If we multiply both sides by $\partial^\mu\phi$ we obtain a nice physical interpretation where the divergence of the stress energy tensor for the scalar field is equal to the force a particle feels in the presence of a wall,

$$\partial_\nu T^{\nu\mu}(\phi) = \sum\int\frac{\mathrm{d}^3k}{(2\pi)^3}f(k)\partial^\mu m^2(\phi). \tag{7.37}$$

The particle distribution functions can be assumed to be a correction to their equilibrium counterparts, $f = f_0 + \delta f$, with $f_0 = (\exp[\beta\gamma(E - vp_z)] \pm 1)^{-1}$ defined in the usual way. The equilibrium terms, f_0, are absorbed into the effective potential and become the one-loop temperature corrections $V(\phi) \to V(\phi, T)$. The particle distribution functions are approximated to obey classical Boltzmann equations

$$\begin{aligned}\left(\frac{\mathrm{d}f}{\mathrm{d}t}\right)_{\text{collisions}} &= \frac{\partial f}{\partial t} + \{f, H\}\\ -C(f) &= \frac{\partial f}{\partial t} + \frac{\partial x}{\partial t}\frac{\partial f}{\partial x} + \frac{\partial p}{\partial t}\frac{\partial f}{\partial p}.\end{aligned} \tag{7.38}$$

Let us calculate the right-hand side explicitly

$$\begin{aligned}-C(f) &= \frac{\partial f}{\partial t} - \left[\frac{\partial x}{\partial t}\frac{\beta\gamma}{E}\frac{\partial m^2}{\partial x} - \frac{\partial p}{\partial t}\beta\gamma\left(\frac{p_x}{E} - v\right)\right]\frac{\mathrm{e}^{\beta\gamma(E - vp_x)}}{\left(\mathrm{e}^{\beta\gamma(E - vp_x)} \pm 1\right)^2}\\ &\equiv \frac{\partial f}{\partial t} - \left[\frac{\partial x}{\partial t}\frac{\beta\gamma}{E}\frac{\partial m^2}{\partial x} - \frac{\partial p}{\partial t}\beta\gamma\left(\frac{p_x}{E} - v\right)\right]h_{F,B}(\beta, E, p_x).\end{aligned} \tag{7.39}$$

If we take our frame of reference to be the bubble wall rest frame then $\delta f = \delta f(x + v_w t)$ with v_w the wall velocity. This implies that $\frac{\partial\delta f}{\partial p} = 0$, $\frac{\partial x}{\partial t} = \frac{p_x}{E}$ and $\frac{\partial p_x}{\partial t} = \frac{-1}{2E}\frac{\partial m^2}{\partial x}v_x$. Furthermore the Boltzmann equations are now reduced to

$$\frac{p_x}{E}\frac{\partial\delta f}{\partial x} - \frac{\partial m^2}{\partial x}\frac{v\beta\gamma}{2E}h_{F,B}(\beta, E, p_z) = -C(f). \tag{7.40}$$

Finally let us define the collision integral. For four-body collisions we sum over all four-legged scattering diagrams evaluated at finite temperature, which means

$$C(f) = \sum \frac{1}{2E_p} \int \frac{\mathrm{d}^3 k \mathrm{d}^3 p' \mathrm{d}^3 k'}{(2\pi)^9 8 E_k E_{p'} E_{k'}} |M(s, t)|^2 (2\pi)^2 \delta(p + k - p' - k') \\ \times \left[f_1 f_2 (1 \pm f_3)(1 \pm f_4) - f_3 f_4 (1 \pm f_1)(1 \pm f_2) \right], \tag{7.41}$$

where the sum of course is over diagrams. We can solve these Boltzmann equations exactly for the case when the collision integral is zero which applies in the case where the mean free path is greater than the bubble wall width,

$$f^{-1} = \exp\left[\frac{E + v_w}{\left(1 - v_w^2\right)T} + s_i \frac{\gamma v_w}{T} \left(\frac{\left(p_x + v_w E\right)^2}{1 - v_w^2} + m_i^2 \right)^{\frac{1}{2}} \right] \pm 1, \tag{7.42}$$

with

$$(m_i, s_i) = \begin{cases} \left(m^2, -1\right) & p_x > -\sqrt{m_0^2 - m^2} \\ \left(m^2 - m_0^2, +1\right) & p_x < -\sqrt{m_0^2 - m^2} \end{cases} \tag{7.43}$$

and the $\pm$ sign in the solution of f^{-1} is for fermions and bosons, respectively. Going beyond this approximation we can consider two regions, the slow and fast wall regimes. Let the collision integral be $C \approx \frac{\delta f}{L_W}$, where L_w is the wall width. In the slow wall regime we assume that $\delta f' \approx \frac{\delta f}{L} \ll \frac{\delta f}{\tau}$ for a diffusion time τ. In this case we can write

$$\delta f = \tau \frac{\partial m^2}{\partial x} \frac{v \beta \gamma}{2E} h_{F,B}(\beta, E, p_x), \tag{7.44}$$

in which case the equations of motion are

$$\frac{\partial^2 \phi}{\partial x^2} + \frac{\partial V(\phi, T)}{\partial \phi} + \sum \frac{\mathrm{d}m^2}{\mathrm{d}\phi} \int \frac{\mathrm{d}^3 p}{(2\pi)^3} \frac{\tau \beta v_w \gamma}{(2E)^2} \frac{\partial m^2}{\partial x} h_{F,B}(\beta, E, p_x) = 0. \tag{7.45}$$

The other region we can assume the fast wall approximation where $\delta f' \approx \frac{\delta f}{L} \gg \frac{\delta f}{\tau}$ we have [24]

$$\frac{p_x}{E} \delta f' = \frac{\partial m^2}{\partial x} \frac{v_w \beta \gamma}{2E} h_{F,B}(\beta, E, p_x) \tag{7.46}$$

in which case the equations of motion are

$$\frac{\partial^2 \phi}{\partial x^2} + \frac{\partial V(\phi, T)}{\partial \phi} + \sum \frac{\mathrm{d}m^2}{\mathrm{d}\phi} \int \frac{\mathrm{d}^3 p}{(2\pi)^3} \frac{m^2 v_w \beta \gamma}{4 E p_x} h_{F,B}(\beta, E, p_x) = 0. \tag{7.47}$$

In both cases the equations have to be solved numerically.

References

[1] Kapusta J I and Landshoff P V 1989 Finite temperature field theory *J. Phys. G: Nucl. Part. Phys.* **15** 267

[2] Kubo R 1957 Statistical-mechanical theory of irreversible processes. I. General theory and simple applications to magnetic and conduction problems *J. Phys. Soc. Japan* **12** 570

[3] Martin P and Schwinger J 1959 Theory of many-particle systems. I. *Phys. Rev.* **115** 1342

[4] Matsubara T 1955 A new approach to quantum-statistical mechanics *Prog. Theor. Phys.* **14** 351

[5] Fick A 1855 Ueber diffusion *Ann. Phys.* **94** 59

[6] Tong D 2012 *Kinetic Theory: University of Cambridge Graduate Course* (Cambridge: University of Cambridge)

[7] Joyce M, Prokopec T and Turok N 1996 Nonlocal electroweak baryogenesis. I. Thin wall regime *Phys. Rev.* D **53** 2930

[8] Joyce M, Prokopec T and Turok N 1995 Electroweak baryogenesis from a classical force *Phys. Rev. Lett.* **75** 9

[9] Arnold P, Moore G D and Yaffe L G 2003 Transport coefficients in high temperature gauge theories, 2. Beyond leading log *J. High Energy Phys.* JHEP05(2003)051

[10] Arnold P, Moore G D and Yaffe L G 2001 Photon emission from quark-gluon plasma: complete leading order results *J. High Energy Phys.* JHEP12(2001)009

[11] Joyce M, Prokopec T and Turok N 1996 Nonlocal electroweak baryogenesis. II. The classical regime *Phys. Rev.* D **53** 6

[12] Kobes R, Kunstatter G and Mak K 1992 Fermion damping in hot gauge theories *Phys. Rev.* D **45** 12

[13] Kobes R L and Semenoff G W 1985 Discontinuities of Green functions in field theory at finite temperature and density *Nucl. Phys.* B **260** 3

[14] Pisarski R D 1989 Scattering amplitudes in hot gauge theories *Phys. Rev. Lett.* **63** 11

[15] Pisarski R D 1993 Damping rates for moving particles in hot, QCD *Phys. Rev.* D **47** 12

[16] Elmfors P *et al* 1999 Damping rates in the MSSM and electroweak baryogenesis *Phys. Lett.* B **452** 3

[17] Kapusta J I and Landshoff P V 1989 Finite-temperature field theory *J. Phys. G: Nucl. Part. Phys.* **15** 3

[18] Chung D J H *et al* 2010 Lepton-mediated electroweak baryogenesis *Phys. Rev.* D **81** 6

[19] Moore G D and Prokopec T 1995 How fast can the wall move? A Study of the electroweak phase transition dynamics *Phys. Rev.* D **52** 7182

[20] John P and Schmidt M G 2001 Do stops slow down electroweak bubble walls? *Nucl. Phys.* B **598** 291

[21] Kozaczuk J 2015 Bubble expansion and the viability of singlet-driven electroweak baryogenesis *J. High Energy Phys.* **10** 1–46

[22] Moore G D and Prokopec T 1995 Bubble wall velocity in a first order electroweak phase transition *Phys. Rev. Lett.* **75** 5
[23] Moore G D 2000 Electroweak bubble wall friction: analytic results *J. High Energy Phys.* JHEP03(2000)006
[24] Huber S J and Sopena M 2013 An efficient approach to electroweak bubble velocities arXiv: hep-ph-1302.1044

A Pedagogical Introduction to Electroweak Baryogenesis

Graham Albert White

Chapter 8

Transport equations

We now have enough information to build a network of transport equations for a given model during a phase transition. In chapter 6 we derived how to relate the divergences of particle number current densities to functions of the particle's self-energies. In the previous chapter we derived the thermal masses and thermal widths which are inputs for the transport coefficients and CP violating sources. We also derived diffusion coefficients and bubble wall velocities as they will be useful in relating the divergence of the current density to derivatives in a single space–time variable of the number densities. In this section we finally put this all together in the MSSM.

8.1 The MSSM under supergauge equilibrium

Gauge and supergauge equilibrium allow us to consider the number densities of gauge singlets summed with their superpartners as a single number density. Furthermore there are some fast interactions that allow us to assume local equilibrium between the two Higgs doublets. If we also ignore all interactions that involve small couplings then our network of transport equations only involve three number densities

$$\begin{aligned} Q &= n_{t_L} + n_{b_L} + n_{\tilde{t}_L} + n_{\tilde{b}_L} \\ T &= n_{t_R} + n_{\tilde{t}_R} \\ H &= n_{H_u^+} + n_{H_u^0} - n_{H_d^-} - n_{H_d^0} \\ & \quad n_{\tilde{H}_u^+} + n_{\tilde{H}_u^0} - n_{\tilde{H}_d^-} - n_{\tilde{H}_d^0}. \end{aligned} \tag{8.1}$$

We can relate chemical potentials to number densities using an assumption of local thermal equilibrium. Ignoring terms of $O(\mu^3)$ we can derive the relation

doi:10.1088/978-1-6817-4457-5ch8

$$\mu_x = \frac{6}{T^2}\frac{n_x}{k_x} \tag{8.2}$$

with

$$k_x = k_x(0)\frac{c_{\mathrm{F,B}}}{\pi^2}\int_{m/T}^{\infty} \mathrm{d}y y \frac{\mathrm{e}^y}{(\mathrm{e}^y \pm 1)^2}\sqrt{y^2 - m_x^2/T^2}, \tag{8.3}$$

where $c_{F(B)} = 6(3)$ and the sign in the denominator is $\pm$ for fermions and bosons, respectively. The factors $k_i(0)$ are 2 for Dirac fermions and complex scalars and 1 for chiral fermions. The k factors of our composite charge densities in equation (8.3) are the sum of the k factors for each component. Combining our number densities together in this way means we also combine the VEV insertion diagrams of tops and stops as well as Higgs and Higgsinos to each form a singlet transport coefficient:

$$\begin{aligned}\Gamma_M^{\pm} &= \Gamma_M^{\pm\,\tilde{t}} + \Gamma_M^{\pm\,t}\\ \Gamma_H^{\pm} &= \Gamma_M^{\pm\,\tilde{H}^{\pm}} + \Gamma_M^{\pm\,\tilde{H}^0}.\end{aligned} \tag{8.4}$$

Furthermore all one-loop three-body rates involving t, Q and H and their superpartners, as cataloged in chapter 6, sum into a single three-body rate, Γ_Y, for the linear combinations of composite densities $\mu_t - \mu_H - \mu_Q$. The set of coupled differential equations is

$$\begin{aligned}\partial_\mu T^\mu = {}& \Gamma_M^{+}\left(\frac{T}{k_T} + \frac{Q}{k_Q}\right) - \Gamma_M^{-}\left(\frac{T}{k_T} - \frac{Q}{k_Q}\right)\\ &- \Gamma_Y\left(\frac{T}{k_T} - \frac{H}{k_H} - \frac{Q}{k_Q}\right)\\ &+ \Gamma_{SS}\left(\frac{2Q}{k_Q} - \frac{T}{k_T} + \frac{9(Q+T)}{k_B}\right) + S_{\tilde{t}}^{C/P}\end{aligned} \tag{8.5}$$

$$\begin{aligned}\partial_\mu Q^\mu = {}& -\Gamma_M^{+}\left(\frac{T}{k_T} + \frac{Q}{k_Q}\right) - \Gamma_M^{+}\left(\frac{T}{k_T} - \frac{Q}{k_Q}\right)\\ &+ \Gamma_Y\left(\frac{T}{k_T} - \frac{H}{k_H} - \frac{Q}{k_Q}\right)\\ &- 2\Gamma_{SS}\left(\frac{2Q}{k_Q} - \frac{T}{k_T} + \frac{9(Q+T)}{k_B}\right) - S_{\tilde{t}}^{C/P}\end{aligned} \tag{8.6}$$

$$\partial_\mu H^\mu = -\Gamma_H\frac{H}{k_H} + \Gamma_Y\left(\frac{T}{k_T} - \frac{Q}{k_Q} - \frac{H}{k_H}\right) + S_{\tilde{H}}^{C/P}, \tag{8.7}$$

where the strong sphaleron rate is taken to be $\Gamma_{SS} \approx 16\alpha_S^4 T$ [1].

8.2 Solution using fast rates, diffusion approximation, and ultrathin wall approximations

Now we have a network of coupled transport equations we will outline the simplest approach to solving them. Unfortunately the approach produces a number for the baryon asymmetry that disagrees with more complete methods by up to two orders of magnitude. Nonetheless it is useful for pedagogical purposes, but we will not go to lengths to justify the approximation we use as they can be viewed as a stepping stone to more complete methods.

From the solution to the network of coupled transport, equations manifest particle–anti-particle asymmetries that diffuse into the symmetric phase ahead of the bubble wall, biasing electroweak sphalerons. The expanding bubble wall will capture some of the net baryon asymmetry. The first step in finding an approximate solution to the network of transport equations is to convert the left-hand sides of the network of transport equations into derivatives of number densities in a single space–time variable. This is achieved by first using the diffusion approximation that we introduced earlier

$$\partial_\mu J^\mu = v_W \dot{n} - D_n \nabla^2 n, \tag{8.8}$$

then making an assumption that the bubble wall is symmetric and we can ignore the curvature of the bubble wall, so that we can reduce the problem to a one-dimensional one. Shifting to the bubble wall rest frame using the variable $z \equiv |v_w t - x|$, we can write the set of coupled transport equations as a set of differential equations in z, whose left-hand sides look like $v_w n_x' - D_x n_x''$. Next we assume both Γ_Y and Γ_{SS} are fast enough such that the linear combination of number densities that these rates are coefficients for are in local equilibrium. That is

$$\begin{aligned} \frac{T}{k_T} - \frac{H}{k_H} - \frac{Q}{k_Q} &= 0 \\ \frac{2Q}{k_Q} - \frac{T}{k_T} + \frac{9(Q+T)}{k_B} &= 0. \end{aligned} \tag{8.9}$$

The second equation can be used to solve T in terms of Q and both can be solved in terms of H using the first equation. The result is

$$\begin{aligned} Q &= \frac{(k_B - 9k_T)k_Q}{\left(9k_T + 9k_Q + k_B\right)k_H} H \\ T &= \frac{(9k_T + 2k_B)k_T}{\left(9k_T + 9k_Q + k_B\right)k_H}. \end{aligned} \tag{8.10}$$

We have reduced our set of transport equations down to a single one. The remaining linearly independent transport equation is any that does not have any fast rate on the right-hand side. One such combination is 2 × (8.5) + (8.6) + (8.7). In this case the transport equation is

$$v_w H'(z) - \overline{D}H''(z) + \overline{\Gamma}H(z) = S(z), \tag{8.11}$$

where

$$\begin{aligned}
\overline{D} &= \frac{\left(9k_Q k_T + k_B k_Q + 4k_T k_B\right)D_q + k_H\left(9k_T + 9k_Q + k_B\right)D_h}{9k_Q k_T + k_B k_Q + 4k_T k_B + k_H\left(9k_Q + 9k_T + k_B\right)} \\
\overline{\Gamma} &= \frac{\left(9k_Q + 9k_T + k_B\right)\left(\Gamma_M^- + \Gamma_H\right) - \left(3k_B + 9k_Q - 9k_T\right)\Gamma_M^+\big)}{9k_Q k_T + k_B k_Q + 4k_T k_B + k_H\left(9k_Q + 9k_T + k_B\right)} \\
S(z) &= \frac{k_H\left(9k_Q + 9k_T + k_B\right)}{9k_T k_Q + k_B k_Q + 4k_T k_B + k_H\left(9k_Q + 9k_T + k_B\right)}\left(S_{\tilde{t}} + S_{\tilde{H}}\right).
\end{aligned} \tag{8.12}$$

Next we make the ultrathin wall approximation where the Higgs profile is assumed to be close to a step function, meaning $S(z)$ and $\overline{\Gamma}$ are zero in the symmetric phase, i.e. when $z > 0$. Furthermore, the functions $\overline{\Gamma}$ become constant in the broken phase in this approximation. This linearizes the remaining transport equations and allows us to solve it in two regions using the method of variable coefficients. The solution in the symmetric phase, up to yet-to-be-determined integration constants that are specified by the boundary and matching conditions, is

$$H(z) = A\mathrm{e}^{v_w z/\overline{D}} + C \tag{8.13}$$

and the solution in the broken phase is

$$\begin{aligned}
H(z) = &\frac{\mathrm{e}^{\kappa_+ z}}{\overline{D}(\kappa_+ - \kappa_-)}\left(\int_0^z \mathrm{d}y\mathrm{e}^{-\kappa_+}S(y) - \beta_+\right) \\
&- \frac{\mathrm{e}^{\kappa_- z}}{\overline{D}(\kappa_+ - \kappa_-)}\left(\int_0^z \mathrm{d}y\mathrm{e}^{-\kappa_-}S(y) - \beta_-\right).
\end{aligned} \tag{8.14}$$

Here the exponentials solve the homogeneous equations and the exponents have the form

$$\kappa_\pm = \frac{v_w \pm \sqrt{v_w^2 + 4\overline{\Gamma}\overline{D}}}{2\overline{D}}. \tag{8.15}$$

We have two boundary conditions in that the densities must go to zero at $\pm\infty$ which sets C to zero and β_+ to

$$\beta_+ = \int_0^\infty \mathrm{d}y\mathrm{e}^{-\kappa_+ y}S(y). \tag{8.16}$$

The other two integration constants are determined by the fact that both $H(z)$ and $H'(z)$ are both continuous at the boundary, $z = 0$. The matching conditions are then

$$A = \frac{\beta_- - \beta_+}{\bar{D}(\kappa_+ - \kappa_-)}$$
$$A(\kappa_+ + \kappa_-) = \frac{\beta_- \kappa_- - \beta_+ \kappa_+}{\bar{D}(\kappa_+ - \kappa_-)}. \tag{8.17}$$

Solving the above equations gives

$$\beta_- = \beta_+ \frac{\kappa_-}{\kappa_+} = \frac{\kappa_-}{\kappa_+} \int_0^\infty \mathrm{d}y e^{-\kappa_+ y} S(y)$$
$$A = -\frac{\beta_+}{\kappa_+} = \frac{1}{\bar{D}\kappa_+} \int_0^\infty \mathrm{d}y e^{-\kappa_+ y} S(y). \tag{8.18}$$

8.3 Solution without fast rates

Here we will now resolve the transport equations without making the assumption that the rates are fast. If we otherwise use the same assumptions as before we once again have set of linearized differential equations which gives us hope for a semi-analytic solution. Note the structure of the transport equations—if we take the combination $\partial_\mu (T + Q)^\mu$ we have a transport equation that depends only on two number densities. If we can create another linear combination that is a function of three number densities only, then the first equation can solve T in terms of Q, the second can solve for H in terms of Q, and the final will have the CP violating source. To create a linear combination of transport equations that is a function of number densities T, Q and H only, we note that the two CP violating sources are proportional to each other

$$S_{\tilde{t}}^{C\!\!/P} = \frac{1}{2a} S_{\tilde{H}}^{C\!\!/P}. \tag{8.19}$$

The set of equations we want is therefore the linear combinations (8.5) + (8.6), $(1 + a) \times (8.5) + (1 - a) \times (8.6) + (8.7)$ and $2 \times (8.5) + (8.6) + (8.7)$.

8.4 Deriving the analytic solution

The problem in deriving an analytic solution quickly becomes messy due to all the k factors and transport coefficients. It is a great simplification to rewrite all transport coefficients in the form a_{Xj}^i, where $j \in \{1, 2, 3\}$ is the equation number, $X \in \{Q, T, H\}$ is the field index, and $i \in \{0, 1, 2\}$ is the power of the derivative in z

$$a_{Q1}^i \partial^i Q + a_{T1}^i \partial^i T = 0 \tag{8.20}$$

$$a_{Q2}^i \partial^i Q + a_{T2}^i \partial^i T + a_{H2}^i \partial^i H = 0 \tag{8.21}$$

$$a^i_{Q3}\partial^i Q + a^i_{T3}\partial^i T + a^i_{H3}\partial^i H = \Delta(z). \tag{8.22}$$

Here of we have used Einstein's summation convention. To solve the first equation for T in terms of Q, use the method of variable coefficients treating Q as an inhomogeneous source. Doing this gives

$$T = \frac{1}{a^2_{T1}}\sum_{\pm}\frac{1}{\kappa_\mp - \kappa_\pm}e^{\kappa_\pm z}\left[\int^z e^{-\kappa_\pm y}\left(a^i_{Q1}\frac{\partial^i Q}{\partial y^i}\right)dy - \beta_i\right]. \tag{8.23}$$

It was demonstrated in [2] that the integration constants are zero. We will ignore them as they clutter notation and little is learned more than the overall counting exercise—we have three number densities and we take their second derivatives so at most we can have six boundary conditions and six matching conditions which specify a total of twelve integration constants. If we substitute the above solution back into the subsequent transport equations we are left with an integro-differential equation. To overcome this we will make a series of variable changes. First,

$$h_\pm = \int^z e^{-\kappa_\pm y} Q dy. \tag{8.24}$$

This eliminates the exponent and integral in equation (8.23), but we now have two functions related via the identity

$$h'_+ = e^{(\kappa_- - \kappa_+)z} h'_-. \tag{8.25}$$

Let us make another change of variables to remove the exponential outside the integral

$$j_\pm = e^{\kappa_\pm z} h_\pm. \tag{8.26}$$

Finally we need to write everything in terms of a single variable. This is achieved via the substitution

$$k = e^{\kappa_\mp z}\int^z e^{-\kappa_\mp y} j_\pm dy, \tag{8.27}$$

which allows us to relate $j_\pm$ to k with the equation

$$j_\pm = k' - \kappa_\mp k. \tag{8.28}$$

We can then relate both T and Q to derivatives in k, which leaves us with differential equations only. It is convenient to rescale k in a way such that

$$T = -a^i_{Q1}\partial^i k \tag{8.29}$$

$$Q = a^i_{T1}\partial^i k. \tag{8.30}$$

We have achieved our initial aim of rewriting the solutions to the differential equation without awkward derivatives or integrals. One can verify by direct substitution that the above indeed satisfies the first transport equation. Next we substitute these equations for T and Q in terms of k into equation (8.21), which gives a differential equation in terms of k and H only,

$$0 = a_{Q2}^{i}\partial^{i}Q + a_{T2}^{i}\partial^{i}T + a_{H2}^{i}\partial^{i}H \tag{8.31}$$

$$= \left(a_{Q2}^{i}a_{T1}^{j} - a_{T2}^{i}a_{Q1}^{j}\right)\partial^{i+j}k + a_{H2}^{i}\partial^{i}H. \tag{8.32}$$

Using the same method as before one finds that the solution for H is given by

$$H = \frac{1}{a_{H2}^{2}}\sum_{\pm}\frac{e^{\kappa_{\pm}z}}{\kappa_{\mp} - \kappa_{\pm}}\int^{z} e^{-\kappa_{\pm}y}\left(a_{Q2}^{i}a_{T1}^{j} - a_{T2}^{i}a_{Q1}^{j}\frac{\partial^{i+j}k}{\partial y^{i}}\right)dy. \tag{8.33}$$

Following the same recipe as before we can write H and k in terms of a variable l

$$H = -\sum_{n=0}^{4}\delta_{i+j-n}\left(a_{Q2}^{i}a_{T1}^{j} - a_{T2}^{i}a_{Q1}^{j}\right)\partial^{n}l \tag{8.34}$$

$$k = a_{H2}^{i}\partial^{i}l. \tag{8.35}$$

In the above we added a Kronecker delta so that the structure of the equations was made more clear. We can then substitute our solution for k into our solutions for T and Q to give all three number densities in terms of l

$$\begin{aligned} H &= -\sum_{n=0}^{4}\delta_{i+j-n}\left(a_{Q2}^{i}a_{T1}^{j} - a_{T2}^{i}a_{Q1}^{j}\right)\partial^{n}l \\ T &= -\sum_{n=0}^{4}\delta_{i+j-n}a_{Q1}^{i}a_{H2}^{j}\partial^{n}l \\ Q &= \sum_{n=0}^{4}\delta_{i+j-n}a_{T1}^{i}a_{H2}^{j}\partial^{n}l. \end{aligned} \tag{8.36}$$

We now can write equation (8.22) in terms of a single variable, l, and a source

$$\begin{aligned} \Delta(z) &= \sum_{n=0}^{6}\delta_{i+j+k-n}\left(a_{T1}^{i}a_{H2}^{j}a_{Q3}^{k} - a_{Q1}^{i}a_{H2}^{j}a_{T3}^{k}\right. \\ &\quad \left. - a_{T1}^{i}a_{Q2}^{j}a_{H3}^{k} + a_{Q1}^{i}a_{T2}^{j}a_{H3}^{k}\right)\partial^{n}l \\ &= \sum_{n=0}^{6}\delta_{i+j+k-n}\epsilon^{abc}a_{Ta}^{i}a_{Hb}^{j}a_{Qc}^{k}\partial^{n}l \\ &\equiv \sum_{n=0}^{6}a_{l}^{n}\partial^{n}l. \end{aligned} \tag{8.37}$$

The step that introduced the permutation symbol makes use of the fact that $a_{H1}^{i} = 0$. Once again using the method of variable coefficients the solution to the above equation is

$$l = \sum_{i=1}^{6} x_i e^{\alpha_i z}\left(\int^{z} e^{-\alpha_i y}\Delta(y)dy - \beta_i\right) \tag{8.38}$$

in the EWSB phase and

$$l = \sum_{i=1}^{6} y_i e^{\gamma_i z} \tag{8.39}$$

in the symmetric phase. In the above we have defined α_i and γ_i as the roots of the polynomials

$$\begin{aligned}\sum_{n=0}^{6} a_l^n \alpha^n = 0 \\ \sum_{n=0}^{6} a_l^n \gamma^n = 0,\end{aligned} \tag{8.40}$$

which are the characteristic polynomials in the broken and symmetric phases, respectively. Furthermore the constants x_i are derived from the equation

$$\vec{x} = M^{-1}\vec{d}. \tag{8.41}$$

In the above the matrix $M_{ij} \equiv \alpha_i^{j-1}$ and j is both an exponent and an index ranging from 1 to 6. Also $\vec{d} \equiv [0, \ldots, 1/a_l^6]^T$. The integration constants are denoted by β_i and y_i in the broken and symmetric phases, respectively. They are determined by matching and boundary conditions. The boundary conditions are that all number densities must be null at $\pm\infty$ whereas the matching conditions are that all number densities and their derivatives must be continuous at the bubble wall. For the symmetric phase the boundary conditions are simple to enforce,

$$y_i = 0 \;\forall \gamma_i \leqslant 0. \tag{8.42}$$

For the broken phase we obtain a condition on the positive exponents

$$x_i\beta_i = x_i \int_0^{\infty} dy e^{-\alpha_i y}\Delta(y) \equiv I_i \;\forall \alpha_i \geqslant 0. \tag{8.43}$$

Finally the matching conditions. First note that we cannot naively match all the derivatives of l at the bubble wall even though doing so indeed gives the correction number of conditions. Rather we match T, Q, and H at the bubble wall along with their first derivatives. Suppose $\gamma_i > 0$ for $i \in \{4, 5, 6\}$ and $\alpha_i > 0$ for $i \in \{1, 2, 3\}$. Let us use the convenient notation that $A_X(\alpha_i) \equiv A_X^{bi}$ and $A_X(\gamma_i) \equiv A_X^{si}$. The matching conditions result in the following conditions

$$\left(x_4\beta_4\ x_5\beta_5\ x_6\beta_6\ x_1\beta_1\ x_2\beta_2\ x_3\beta_3\ y_1\ y_2\ y_3\right)^T =$$
$$= \begin{pmatrix} 1 & 0 & 0 & 0 & 0 & 0 & 0 & 0 & 0 \\ 0 & 1 & 0 & 0 & 0 & 0 & 0 & 0 & 0 \\ 0 & 0 & 1 & 0 & 0 & 0 & 0 & 0 & 0 \\ A_Q^{b4} & A_Q^{b5} & A_Q^{b6} & A_Q^{b1} & A_Q^{b2} & A_Q^{b3} & A_Q^{s1} & A_Q^{s2} & A_Q^{s3} \\ \alpha_4 A_Q^{b4} & \alpha_5 A_Q^{b5} & \alpha_6 A_Q^{b6} & \alpha_1 A_Q^{b1} & \alpha_2 A_Q^{b2} & \alpha_3 A_Q^{b3} & \gamma_1 A_Q^{s1} & \gamma_2 A_Q^{s2} & \gamma_3 A_Q^{s3} \\ A_T^{b4} & A_T^{b5} & A_T^{b6} & A_T^{b1} & A_T^{b2} & A_T^{b3} & A_T^{s1} & A_T^{s2} & A_T^{s3} \\ \alpha_4 A_T^{b4} & \alpha_5 A_T^{b5} & \alpha_6 A_T^{b6} & \alpha_1 A_T^{b1} & \alpha_2 A_T^{b2} & \alpha_3 A_T^{b3} & \gamma_1 A_T^{s1} & \gamma_2 A_T^{s2} & \gamma_3 A_T^{s3} \\ A_H^{b4} & A_H^{b5} & A_H^{b6} & A_H^{b1} & A_H^{b2} & A_H^{b3} & A_H^{s1} & A_H^{s2} & A_H^{s3} \\ \alpha_4 A_H^{b4} & \alpha_5 A_H^{b5} & \alpha_6 A_H^{b6} & \alpha_1 A_H^{b1} & \alpha_2 A_H^{b2} & \alpha_3 A_H^{b3} & \gamma_1 A_H^{s1} & \gamma_2 A_H^{s2} & \gamma_3 A_H^{s3} \end{pmatrix}^{-1} \begin{pmatrix} I_1 \\ I_2 \\ I_3 \\ 0 \\ 0 \\ 0 \\ 0 \\ 0 \\ 0 \end{pmatrix} \tag{8.44}$$
$$\equiv \begin{pmatrix} \mathbf{1}_{3\times 3} & 0 \\ \vec{A}_X(\alpha) & \vec{A}_X(\gamma) \\ (\vec{\alpha A})_X(\alpha) & (\vec{\alpha A})_X(\gamma) \end{pmatrix}^{-1} \begin{pmatrix} I_1 \\ I_2 \\ I_3 \\ 0 \end{pmatrix}.$$

We can now write the analytic form of the solutions for H, T, and Q:

$$\begin{aligned} H &= \sum_{i=1}^{6} A_H(\alpha_i) x_i e^{\alpha_i z}\left(\int^z e^{-\alpha_i y}\Delta(y)dy - \beta_i\right) \\ T &= \sum_{i=1}^{6} A_T(\alpha_i) x_i e^{\alpha_i z}\left(\int^z e^{-\alpha_i y}\Delta(y)dy - \beta_i\right) \\ Q &= \sum_{i=1}^{6} A_Q(\alpha_i) x_i e^{\alpha_i z}\left(\int^z e^{-\alpha_i y}\Delta(y)dy - \beta_i\right), \end{aligned} \tag{8.45}$$

with known functions defined as

$$\begin{aligned} A_H &= -\sum_{n=0}^{4} \delta_{i+j-n}\left(a_{Q2}^i a_{T1}^j - a_{T2}^i a_{Q1}^j\right)\alpha^n \\ A_T &= -\sum_{n=0}^{4} \delta_{i+j-n} a_{Q1}^i a_{H2}^j \alpha^n \\ A_Q &= \sum_{n=0}^{4} \delta_{i+j-n} a_{T1}^i a_{H2}^j \alpha^n. \end{aligned} \tag{8.46}$$

So we have derived a fairly simple solution to the transport equations for the MSSM case. However, this calculation is easily generalized to multiple transport equations. Also it is significantly simpler to assume the form of the solutions, parametrized in terms of α, $A_X(\alpha)$, β, and x_i, and substitute back into the transport equations to obtain a set of conditions which specify the above parameters. Specifically α and

$A_X(\alpha)$ solve the homogeneous version of the transport equations and x_i is defined when substituting the full solution into the inhomogeneous transport equations. Furthermore it is straightforward to deal with the case of multiple transport equations—simply replace x_i with x_{ij} and the sources $\Delta(z)$ become replaced with $\Delta_j(z)$. One can then proceed as described above, either deriving the solution or substituting the known form to derive the parameters.

8.5 Beyond ultrathin walls

In section 4.6 a method was introduced for solving second-order differential equations by approximating them with a linear set of equations and then perturbing around the solution to that. When we approximate the mass relaxation terms as constant in the broken phase and zero in the other, we are linearizing our differential equations. Therefore we can also write perturbations to the solution which take into account the space–time dependence of the mass terms. Consider the second transport equation (8.22), which contains mass terms proportional to a. Let us assume for simplicity that a is small so that the only mass terms are in equation (8.22). We will define the following error functions

$$\begin{aligned} \Delta(z) &= \Theta(z)\Delta(z) + (1 - \Theta(z))\Delta(z) \equiv \Delta_0(z) + \epsilon(z) \\ a_{l3}^0(z) &\equiv a_{l3}^0(z)\Theta(z) + a_{l3}^0(z)\Theta(-z) \\ &= a_{l3}^0(z_{\max}) - \left[a_{l3}^0(z_{\max}) - a_{l3}^0(z)\right] + a_{l3}^0(z)\Theta(-z) \\ &= a_{l3}^0 + \delta a_{l3}^0(z) \\ l(z) &= l_0(z) + \delta_1 l(z) + \delta_2 l(z) + \cdots. \end{aligned} \tag{8.47}$$

Note that $l_0(z)$ solves the linearized transport equations so we can make a dramatic simplification similar to the simplification we made when solving bubble wall profiles. Also, just like in the case of the bubble wall profiles, these perturbations are finite which allows for a perturbative series. Inserting the perturbations into the transport equations gives

$$\begin{aligned} &a_{l3}^i \partial^i l_0 + a_{l3}^i \partial i\left(\delta_1 l + \delta_2 l + \cdots\right) + \delta a_{l3}^0(z)\left(l_0 + \delta_1 l + \delta_2 l + \cdots\right) \\ &= \Delta(z) + \epsilon(z). \end{aligned} \tag{8.48}$$

We can use the exact same method as before to find the corrections to m

$$\begin{aligned} \delta_1 l &= \sum_{i=0}^{6} \mathrm{e}^{\alpha_i z} x_i \left(\int^z \mathrm{e}^{-\alpha_i y}\left[\epsilon - l_0 \delta a_{l3}^0(z)\right] - \delta_1 \beta_i\right) \\ \delta_2 l &= \sum_{i=0}^{6} \mathrm{e}^{\alpha_i z} x_i \left(\int^z \mathrm{e}^{-\alpha_i y}\left[- \delta_1 l \delta a_{l3}^0(z)\right] - \delta_2 \beta_i\right), \end{aligned} \tag{8.49}$$

etc.

References

[1] Moore G D 2000 Sphaleron rate in the symmetric electroweak phase *Phys. Rev.* D **62** 8
[2] White G A 2016 General analytic methods for solving coupled transport equations: from cosmology to beyond *Phys. Rev.* D **93** 4

Chapter 9

The baryon asymmetry

Finally let us perform the calculation, then put everything together and use the solutions to our transport equations to calculate the baryon asymmetry of the Universe. Weak sphalerons interact with the total left-handed number density to produce a baryon asymmetry. From the solutions to the transport equations, one can then define the left-handed number density $n_L(z) = Q_{1L} + Q_{2L} + Q_{3L} = 5Q + 4T$. The baryon number density, ρ_B, satisfies the equation for a certain relaxation constant $\mathcal{R}$ [1, 2]

$$D_Q\rho_B''(z) - v_W\rho_B'(z) - \Theta(-z)\mathcal{R}\rho_B = \Theta(-z)\frac{n_F}{2}\Gamma_{ws}n_L(z), \tag{9.1}$$

where n_F is the number of fermion families. The relaxation parameter can be derived by asserting that

$$A\frac{n_B}{T^2} = \mu_{\rm CS} = 9\mu_q + \mu_{l_i}, \tag{9.2}$$

where $\mu_{\rm CS}$ is the chemical potential for the Chern–Simons number for the $SU(2)_L$ gauge field. The second equation in the above comes from the fact that a weak sphaleron creates nine quarks and three leptons. We can assume quark mixing interactions are fast compared to the weak sphaleron rate in the symmetric phase. We can then write for the Standard Model

$$n_B = \frac{1}{3}n_q \to \mu_q = \frac{1}{2}\frac{\mu_q}{T^2} \tag{9.3}$$

and the lepton number can be related to the baryon number via $\sum_i\mu_{l_i} = 2n_B/T^2$ under the assumption that the right-handed leptons do not have time to equilibrate which gives $A = 15/2$. For the MSSM a simple calculation reveals

$$\mathcal{R} = \Gamma_{ws}\left[\frac{9}{4}\left(1 + \frac{n_{sq}}{6}\right) + \frac{3}{2}\right] \tag{9.4}$$

where n_{sq} is the number of squarks thermally available at the nucleation temperature and $\Gamma_{ws} \approx 120\alpha_W^5 T$ [3]. The baryon asymmetry of the Universe, Y_B is then given by

$$Y_B = -\frac{n_F \Gamma_{ws}}{2\kappa_+ D_Q S} \int_{-\infty}^{0} \mathrm{e}^{-\kappa_- x} n_L(x) \mathrm{d}x, \tag{9.5}$$

where

$$\kappa_\pm = \frac{v_W \pm \sqrt{v_W^2 + 4D_Q \mathcal{R}}}{2D_Q} \tag{9.6}$$

and the entropy is given by

$$S = \frac{2\pi^2}{45} g_* T^3. \tag{9.7}$$

References

[1] Moreno J M, Quiros M and Seco M 1998 Bubbles in the supersymmetric standard model *Nucl. Phys.* B **526** 489–500

[2] Cline J M, Joyce M and Kainulainen K 2000 Supersymmetric electroweak baryogenesis *J. High Energy Phys.* JHEP07(2000)018

[3] Bödeker D, Moore G D and Rummukainen K 2000 Chern-Simons number diffusion and hard thermal loops on the lattice *Phys. Rev.* D **61** 5

Graham Albert White

Chapter 10

A brief phenomenological summary

In this section we provide a brief review of the phenomenological consequences and limits of electroweak baryogenesis in supersymmetry (SUSY) models with a particular emphasis on the MSSM. Unfortunately, baryogenesis within the MSSM is all but ruled out apart from tiny corners of the parameter space [1, 2]. However, the lessons than can be learned from the MSSM are a useful guide for examining extensions to the MSSM.

The first lesson from the MSSM is that more areas of the parameter space become unphysical when one considers our cosmic history. If squarks acquire a vacuum expectation value at some temperature, it has been shown that the tunneling rate from the colour breaking phase to the SM phase is too feeble for the transition to occur [3]. Therefore it is not enough to ensure that the SM vacuum is the deepest minimum, but it must also either be the deepest minimum at all temperatures or the tunneling rate to the colour breaking minimum must be sufficiently suppressed. This problem is particularly pronounced in the MSSM as the stop mass-squared parameter must be negative to increase the strength of the electroweak phase transition. Apart from the phenomenological problems that this causes in making one of the stop masses light, a negative mass-squared parameter creates a minimum in a colour breaking direction.

To catalyze a strongly first-order phase transition, new scalar particles at around the weak scale are needed. New scalars can catalyze such a phase transition either by [4–9]

- Effective thermal cubic corrections from the $[m^2(\phi)]^{3/2}$ terms that arise in the high temperature expansion. Only bosons have a cubic term in their high temperature expansion, which is why new fermions cannot do this job.
- The scalar can acquire a vacuum expectation value before the EWPT and either change the angle of the transition in field space—a direction which contains a barrier separating the transition—or delay the transition such that the critical temperature is lower.

doi:10.1088/978-1-6817-4457-5ch10

Surprisingly, these scalars can be up to around a TeV in mass and still catalyze a strongly first-order phase transition through the second mechanism—that is, so long as the extra scalar is not a squark since the squark can never obtain a vacuum expectation value. The effect of the first mechanism is boosted if the scalar is coloured as you obtain a factor of N_C. If the scalar is a gauge singlet then large Higgs portal couplings are required for the first mechanism to work.

For the MSSM the squark must only contribute through the first mechanism but strong LHC bounds on squark masses makes such a scenario virtually impossible. Extensions of the MSSM can contain new scalars such as gauge singlets in the NMSSM which can easily catalyze a strongly first-order electroweak phase transition [10–12]. The main constraints on gauge singlets are from electroweak precision constraints—the mixing angle in the scalar mass matrix cannot be too large. Also, in general, the larger the singlet scalar mass the larger the mixing angle needs to be in order to catalyze a strongly first-order electroweak phase transition. A pitfall though in singlet extensions is that the strength of the EWPT can be so large that tunneling rates become overly suppressed and electroweak symmetry breaking may not proceed at all. Even if one boosts the strength of the phase transition, some care needs to be taken in checking exactly how strong the order parameter needs to be, as it varies from model to model. For example, it was shown that in the two-Higgs doublet model the order parameter $\phi_c/T_c > 1.2$ [13, 14].

The next phenomenological concern is the constraints imposed by searches for permanent EDMs. Such experiments are arguably providing the stringent constraints, particularly if your CP violation is the stop where the Yukawa coupling is close to unity. One can suppress the one-loop contributions to EDMs by observing that all diagrams that contribute at one loop contains one scalar (squark or slepton) and one fermion (Higgsino or gaugino). If one makes one of the species heavy then the one loop EDMs become suppressed.

In general this is not always enough. If your CPV phase contributes to two-loop Bar–Zee EDM diagrams, then the current constraints from negative searches for permanent EDMs imply that the masses of the particles involved need to be quite heavy. This makes it difficult to drive enough production of asymmetry unless the CPV phase is small to compensate, but this means one has to be very near a resonance (that is near degenerate masses) to produce enough BAU [15]. In the MSSM one is left relying on neutralino–Higgsino interactions and the regions of parameter space left to hide are critically endangered if not extinct. Gauge singlet extensions of the MSSM appear to be able to still successfully drive electroweak baryogenesis.

A potentially useful feature of gauge singlets is that the variation of the ratio of Higgs VEVs, $\Delta\beta$, can be an order of magnitude higher than in models with just two Higgs doublets [16]. Since the baryon asymmetry is generically proportional to $\Delta\beta$, this provides an opportunity for either off-resonance baryogenesis or baryogenesis being produced with small CPV phases. One positive for SUSY models in general is that coloured scalars provide a very strong source of friction for bubble walls suppressing their velocity. In general baryogenesis prefers a small wall velocity so that the diffusion time is sufficiently suppressed, although there has been some recent work on the effect of fast bubble walls [17]. One, however, has to be cautious to check in an arbitrary model that the wall velocity is not so small that diffusion time becomes small compared to the weak sphaleron rate, as such a scenario would cause the baryon asymmetry to vanish.

Finally the recent discovery of gravitational waves makes the gravitational wave background produced by cosmic phase transitions an attractive area of study (see for instance [18–26]). Background gravitational waves are created both in the expansion of the bubble walls and the collision of bubbles. Unfortunately the closest gravitational wave experiment on the horizon, LIGO, does not have a sensitivity to the background that would be produced by a phase transition at the electroweak scale. On the other hand, they do have a sensitivity to phase transitions occurring at temperatures $\approx 10^7$–10^8 GeV [27]. Since EDM, electroweak precision and collider constraints continue to squeeze the parameter space for vanilla electroweak baryogenesis; the non-minimal framework of a multi-step phase transition has the attractive feature of being potentially corroborated at LIGO if the crucial phase transition occurs at a fortuitous temperature. Further in the future, LISA promises to be sensitive to the background gravitational waves caused during a strongly first-order phase transition, which could provide a tantalizing confirmation of one of the key ingredients of electroweak baryogenesis [28].

References

[1] Carena M *et al* 2009 The baryogenesis window in the MSSM *Nucl. Phys.* B **812** 1

[2] Curtin D, Jaiswal P and Meade P 2012 Excluding electroweak baryogenesis in the MSSM *J. High Energy Phys.* **8** 1–27

[3] Cline J M, Moore G D and Servant G 1999 Was the electroweak phase transition preceded by a color-broken phase? *Phys. Rev.* D **60** 10

[4] Profumo S, Ramsey-Musolf M J and Shaughnessy G 2007 Singlet Higgs phenomenology and the electroweak phase transition *J. High Energy Phys.* JHEP08(2007)010

[5] Profumo S *et al* 2015 Singlet-catalyzed electroweak phase transitions and precision Higgs boson studies *Phys. Rev.* D **91** 3

[6] Laine M and Rummukainen K 1998 The MSSM electroweak phase transition on the lattice *Nucl. Phys.* B **535** 1

[7] Espinosa J R, Quiros M and Zwirner F 1993 On the electroweak phase transition in the minimal supersymmetric Standard Model *Phys. Lett.* B **307** 1–2

[8] Farrar G R and Losada M 1997 SUSY and the electroweak phase transition *Phys. Lett.* B **406** 1

[9] Brignole A *et al* 1994 Aspects of the electroweak phase transition in the minimal supersymmetric Standard Model *Phys. Lett.* B **324** 2

[10] Huber S J *et al* 2006 Electroweak phase transition and baryogenesis in the NMSSM *Nucl. Phys.* B **757** 1

[11] Carena M, Shah N R and Wagner C E M 2012 Light dark matter and the electroweak phase transition in the NMSSM *Phys. Rev.* D **85** 3

[12] Funakubo K, So Tao and Toyoda F 2005 Phase transitions in the NMSSM *Prog. Theor. Phys.* **114** 2

[13] Fuyuto K and Senaha E 2014 Improved sphaleron decoupling condition and the Higgs coupling constants in the real singlet-extended Standard Model *Phys. Rev.* D **90** 015015
[14] Fuyuto K and Senaha E 2015 Sphaleron and critical bubble in the scale invariant two Higgs doublet model *Phys. Lett.* B **747** 152
[15] Morrissey D E and Ramsey-Musolf M J 2010 Electroweak baryogenesis *New J. Phys.* **14** 125003
[16] Kozaczuk J *et al* 2015 Cosmological phase transitions and their properties in the NMSSM *J. High Energy Phys.* JHEP01(2015)144
[17] Caprini C and No J M 2012 Supersonic electroweak baryogenesis: achieving baryogenesis for fast bubble walls *J. Cosmol. Astropart. Phys.* JCAP01(2012)031
[18] Apreda R *et al* 2002 Gravitational waves from electroweak phase transitions *Nucl. Phys.* B **631** 1
[19] Grojean C and Servant G 2007 Gravitational waves from phase transitions at the electroweak scale and beyond *Phys. Rev.* D **75** 4
[20] Kosowsky A, Turner M S and Watkins R 1992 Gravitational waves from first-order cosmological phase transitions *Phys. Rev. Lett.* **69** 14
[21] Binétruy P *et al* 2012 Cosmological backgrounds of gravitational waves and eLISA/NGO: phase transitions, cosmic strings and other sources *J. Cosmol. Astropart. Phys.* JCAP06 (2012)027
[22] Kamionkowski M, Kosowsky A and Turner M S 1994 Gravitational radiation from first-order phase transitions *Phys. Rev.* D **49** 6
[23] Linde A D 1979 Phase transitions in gauge theories and cosmology *Rep. Prog. Phys.* **42** 3
[24] Nicolis A 2004 Relic gravitational waves from colliding bubbles and cosmic turbulence *Class. Quantum Grav.* **21** 4
[25] Huber S J and Konstandin T 2008 Gravitational wave production by collisions: more bubbles *J. Cosmol. Astropart. Phys.* JCAP09(2008)022
[26] Mark Hindmarsh *et al* 2014 Gravitational waves from the sound of a first order phase transition *Phys. Rev. Lett.* **112** 4
[27] Dev P S B and Mazumdar A 2016 Probing the scale of new physics by Advanced LIGO/VIRGO *Phys. Rev.* D **93** 104001
[28] Klein A *et al* 2016 Science with the space-based interferometer eLISA: supermassive black hole binaries *Phys. Rev.* D **93** 024003

Chapter 11

Other mechanisms for producing the baryon asymmetry

Here we have reviewed the vanilla electroweak baryogenesis framework with an emphasis on the theoretical background. Many other mechanisms have been proposed for producing the BAU (for a recent example in the electroweak framework or thereabouts see [1–12] and [13–18] for recent examples outside the electroweak baryogenesis framework). In this section we briefly review three other prominent paradigms for producing the BAU

- *The Affleck–Dine model and leptogenesis.* These models belong to a different paradigm to making use of phase transitions. In the case of leptogenesis the departure from equilibrium occurs when the temperature drops below the mass of a heavy particle, which decays into lighter particles through lepton violating interactions. In the case of the Affleck–Dine mechanism, scalar particles carrying baryon number interact with the inflaton during the reheating period after inflation. Once again the heavy particle decays into leptons and baryons.
- *The Peskin–Anderson mechanism.* Another paradigm to the above, this model utilizes cosmic inflation itself as the departure from equilibrium. Since inflation generally washes out any baryon asymmetry, making use of it—and thereby turning an enemy into a friend—was something of an interesting paradigm shift. It makes use of the gravitational anomaly to violate baryon number conservation.

Unfortunately not all of these models will be testable in the foreseeable future, but they are all theoretically elegant and are a demonstration of the variety of answers there might be to the question of how the asymmetry between particles and antiparticles arose.

doi:10.1088/978-1-6817-4457-5ch11

11.1 Leptogenesis

The only particle experiment to convincingly show a departure from the Standard Model is the detection of nuetrino masses [19]. Their mass appears to be unnaturally small compared to other Standard Model particles. A mechanism for producing their mass as well as naturally ensuring that they are small is the see-saw mechanism. This mechanism works by adding a Majorana mass to the Standard Model, which is allowed since the right-handed neutrino is a gauge singlet [20]:

$$\mathcal{L}_\nu \ni y_e^{ij}\bar{e}_i L_j H^\dagger + y_\nu^{ij}\bar{\nu}_i L_j H - \frac{M_i}{2}\nu_i\nu_i. \tag{11.1}$$

The Lagrangian mass parameter M_i is a free parameter and it is expected to be GUT scale. The two mass eigenstates involve a heavy one around the scale of M and one suppressed by v^2/M. With order 1 couplings and a GUT scale of 10^{15} GeV one obtains a light neutrino mass of about 0.01 eV.

Apart from providing a natural mechanism for small neutrino masses, these mass terms in the Lagrangian violate the lepton number explicitly. When the temperature of the Universe is larger than M there is an abundance of right-handed neutrinos. The departure from equilibrium occurs when the temperature of the Universe drops below M and these particles decay into Higgs and left-handed leptons. The Yukawa couplings between the Higgs and the neutrino are complex and are the source of CP violation as the tree-level decay interferes with the one-loop decay, as shown in figure 11.1. The difference between the decay to leptons and anti-leptons is denoted by epsilon [21–23]

$$\epsilon = \frac{\Gamma(\nu \to H + l) - \Gamma(\nu \to H^\dagger + l^\dagger)}{\Gamma(\nu \to H + l) + \Gamma(\nu \to H^\dagger + l^\dagger)}. \tag{11.2}$$

For a similar mass hierarchy to the quark hierarchy one finds that $\epsilon \in [10^{-5}, 10^{-6}]$.

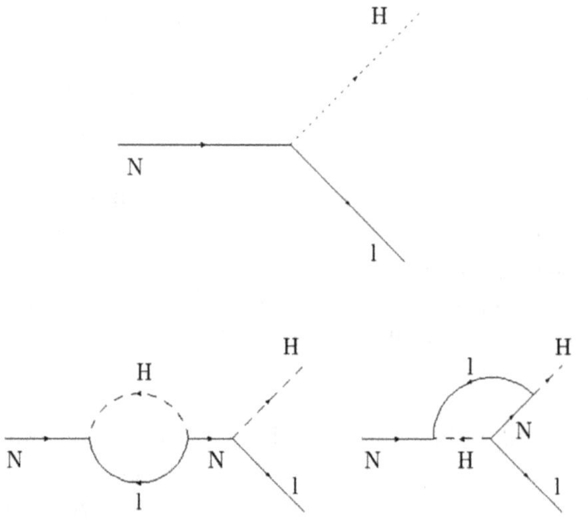

Figure 11.1. Tree and one-loop decay of the heavy neutrino. The CP violating phases governing the process ensures that the interference between tree and one-loop effects will lead to an asymmetry between leptons and anti-leptons.

Some of the lepton asymmetry is then passed to the baryon sector through weak sphaleron interactions. To see how such a transfer takes place we make use of the fact that weak sphalerons violate $B + L$ but conserve $B - L$. If one assumes all Standard Model Yukawa and gauge interactions are fast enough to set the combination of chemical potentials involved in each subsequent reaction in equilibrium, one has the following expression for the baryon and lepton numbers in terms of Standard Model particles:

$$\begin{aligned}\mu_B &= \sum_i 2\mu_{Q_i} + \mu_{u_i} + \mu_{d_i} \\ \mu_L &= \sum_i 2\mu_{L_i} + \mu_{e_i}.\end{aligned} \tag{11.3}$$

If we apply the conservation of hypercharge and set the flavor changing interactions to equilibrium, one finds that

$$\begin{aligned}B &= \frac{28}{79}(B - L) \\ L &= \frac{51}{79}(B - L).\end{aligned} \tag{11.4}$$

We can therefore calculate the baryon asymmetry from the lepton asymmetry

$$Y_B = \frac{n_B}{S} \approx -\frac{28}{79}\frac{n_L}{S} \approx 10^{-3}\eta\epsilon, \tag{11.5}$$

where η is an efficiency factor that comes from solving the set of Boltzman equations. For $\eta\epsilon \approx 10^{-7}$ one obtains the correct baryon asymmetry.

Leptogenesis is very simple and elegant but it is not without problems. The large Majorana mass causes a very severe fine tuning problem (this of course is the usual hierarchy problem). If one solves such a hierarchy problem with SUSY then one runs into the problem that without some fine-tuning there is an overabundance of gravitinos. The biggest problem with leptogenesis, however, at least from the perspective of phenomenologists, is that the particles involved are so heavy (or alternatively so weakly coupled) that the scenario is untestable in the near future. One can indirectly corroborate it by tests of neutrinoless double beta decay which could, in principle, establish the Majorana nature of neutrinos. Assuming such an endeavor is successful, constraints on viable leptogenesis models can come through a measurement of the lepton asymmetry of the Universe (and comparing it to the BAU), as well as cosmic constraints on the neutrino mass sum as well as the masses of light neutrinos.

11.2 Affleck–Dine

The Affleck–Dine mechanism is similar to the Leptogenesis approach in that it involves decays of heavy particles [24, 25]. In this case though the mechanism is a little more subtle, hence giving it separate treatment. During inflation, scalar fields with a shallow potential can acquire vacuum expectation values. SUSY, for example, has many such directions which can be scalar fields, ϕ, with non-zero lepton number [28]. It is possible that such a field could have a large VEV during inflation and both C and CP are spontaneously violated when this happens. The large Hubble parameter would act as a very strong friction term that prevents the condensate from decaying during inflation. After inflation the field rotates in field space and the condensate decays into fermions with non-zero lepton number. Consider an effective potential of the form

$$V(\phi) = \left(m_\phi^2 + c_H H^2\right)\phi^2 + O\left(\phi^4\right) + \cdots. \tag{11.6}$$

We have just written the mass term due to it being the term of most interest for this discussion. Consider the case where $c_H < 0$. The Hubble parameter is very large during inflation and the field will have a large vacuum expectation value. Once inflation stops the Hubble parameter decreases until the minimum of the potential is at the origin and the condensate will decay. The equation of motion for ϕ is

$$\ddot{\phi} + 3H\dot{\phi} + \frac{\partial V}{\partial \phi^*} = 0. \tag{11.7}$$

Since ϕ has lepton number one can write the lepton number density as

$$n_L = \mathrm{i}\beta\left(\dot{\phi}\phi^* - \phi\dot{\phi}^*\right). \tag{11.8}$$

Using the above two equations we have

$$\dot{n}_L + 3Hn_L + 2\beta\,\mathrm{Im}\left[\frac{\partial V}{\partial \phi^*}\phi^*\right]. \tag{11.9}$$

The field during reheating spirals into the origin, which means a non-zero lepton number throughout. This can decay into a permanent lepton number due to couplings between the scalar field and Standard Model leptons. The lepton asymmetry can then be transferred to a baryon asymmetry through electroweak sphalerons.

11.3 Using inflation

How to complete the Standard Model to include gravity in a UV safe way is an open problem. However, one can just use canonical quantum gravity as an effective field theory at low energy. Gravitons then become gauge bosons that mediate gravitational interactions and the corresponding gauge invariance is local Lorentz symmetry. Taking the triangle diagram of figure 3 and replacing the SM gauge bosons with gravitons, one derives an anomalous current. Using the Fujikawa method it is straightforward to derive the gravitational anomaly [26]

$$\partial_\mu J_L^\mu = R\tilde{R}, \tag{11.10}$$

where the current J_L^μ is the anomalous lepton current. The departure from equilibrium occurs through inflation itself, which is what makes this mechanism paradigm shifting—it turns an enemy into a friend. Suppose the inflaton is an axion field. The inflaton will couple to the anomalous lepton number current gravitationally via the interaction $F(\phi)R\tilde{R}$, where F is an odd function in ϕ. We can write the metric of the Universe with tensor fluctuations as

$$ds^2 = -dt^2 + a^2(t)\left(dx^2 + h_{ij}dx^i dx^j\right). \tag{11.11}$$

Gauge invariance gives us the freedom such that only two transverse traceless elements of h_{ij} are non-zero, which we denote h_L and h_R,

$$\begin{aligned} ds^2 = -dt^2 + a^2(t)\Bigg(&\left[1 - \frac{1}{\sqrt{2}}(h_L + h_R)\right]dx^2 \\ &+\left[1 + \frac{1}{\sqrt{2}}(h_L + h_R)\right]dy^2 + i\sqrt{2}(h_R - h_L)dxdy + dz^2\Bigg). \end{aligned} \tag{11.12}$$

The anomalous lepton current can then be written explicitly by calculating the Ricci tensor from the above metric and contracting it with its dual

$$R\tilde{R} = \frac{4i}{a^3}\left[\partial_z^2 h_r \partial_z \partial_t h_L + a^2 \partial_t^2 h_R \partial_t \partial_z h_L + \frac{1}{2}\partial_t a^2 \partial_t h_R \partial_t \partial_L h_L - (L \leftrightarrow R)\right]. \tag{11.13}$$

From the above it is obvious that the anomalous lepton number current vanishes in the case of left–right symmetry. We need the gravitational waves to have some birefringence. This is achieved by adding the term $F(\phi)R\tilde{R}$ to the Einstein–Hilbert action. Choosing a simple form for [27]

$$F = \frac{N\phi}{16\pi^2 M_{\rm pl}^2}, \tag{11.14}$$

where $m_{\rm pl}$ is the Planck mass and N is a dimensionless number motivated by string theory, the classical equations of motion give

$$\begin{aligned} \Box h_L &= -2i\frac{\Theta}{a}\dot{h}_L' \\ \Box h_R &= 2i\frac{\Theta}{a}\dot{h}_R', \end{aligned} \tag{11.15}$$

with

$$\Theta = \frac{N\dot{\phi}H}{2\pi^2 M_{\rm pl}^3}, \tag{11.16}$$

which indeed has a birefringence while inflation is happening. In fact one can rewrite the above equation in terms of the slow roll parameter $\epsilon = \frac{1}{2}\dot{\phi}^2/(HM_{\rm pl})^2$

$$\Theta = \sqrt{2\epsilon}N\frac{H^2}{2\pi^2 M_{\rm pl}^2}. \tag{11.17}$$

This elegant and relatively new idea is unfortunately well outside experimental reach, unless there is a dramatic improvement in the precision of searches for non-Gaussianities in the CMB radiation which gives us sufficiently detailed information about the inflaton.

References

[1] Balázs C *et al* 2014 Baryogenesis, dark matter and inflation in the next-to-minimal supersymmetric standard model. *J. High. Energy Phys.* arXiv:1309.5091(2013)

[2] Alanne T *et al* 2016 Baryogenesis in the two doublet and inert singlet extension of the Standard Model *J. Cosmol. Astropart. Phys.* JCAP08 (2016)057

[3] Chiang C-W, Kaori F and Eibun S 2016 Electroweak baryogenesis with lepton flavor violation *Phys. Lett. B at press* http://www.sciencedirect.com/science/article/pii/S0370269316305573

[4] Phong V Q, Thao N C and Long H N 2015 Baryogenesis in the Zee-Babu model arXiv: hep-ph-1511.00579

[5] Rindler-Daller T, Li B, Shapiro P R, Lewicki M and Wells J D 2015 How scalar-field dark matter may conspire to facilitate baryogenesis at the electroweak scale *Mtg of the APS Division of Particles and Fields (DPF 2015) (Ann Arbor, MI, August 4–8)* arXiv: hep-ph-1510.08369

[6] Arcadi G, Covi L and Nardecchia M 2015 Gravitino Dark Matter and low-scale baryogenesis *Phys. Rev.* D **92** 115006

[7] Chao W 2015 Electroweak Baryogenesis in the Exceptional Supersymmetric Standard Model *J. Cosmol. Astropart. Phys.* JCAP08(2015)055

[8] Dorsch G C, Huber S J, Mimasu K and No J M 2014 Echoes of the electroweak phase transition: discovering a second Higgs doublet through $A_0 \rightarrow ZH_0$ *Phys. Rev. Lett.* **113** 211802

[9] Chang X and Huo R 2014 Electroweak baryogenesis in the MSSM with vectorlike superfields *Phys. Rev.* D **89** 036005

[10] Lewicki M, Rindler-Daller T and Wells J D 2016 Enabling electroweak baryogenesis through dark matter *J. High. Energy Phys.* JHEP06(2016)055

[11] Dhuria M, Hati C and Sarkar U 2016 Moduli induced cogenesis of baryon asymmetry and dark matter *Phys. Lett.* B **756** 376

[12] Fuyuto K, Hisano J and Senaha E 2016 Toward verification of electroweak baryogenesis by electric dipole moments *Phys. Lett.* B **755** 491

[13] Ipek S and March-Russell J 2016 Baryogenesis via particle-antiparticle oscillations *Phys. Rev.* D **93** 123528

[14] Fukushima M, Mizuno S and Maeda K-i 2016 Gravitational Baryogenesis after anisotropic inflation *Phys. Rev.* D **93** 103513

[15] Fujita T and Kamada K 2016 Large-scale magnetic fields can explain the baryon asymmetry of the Universe *Phys. Rev.* D **93** 083520

[16] Yamada M 2016 Affleck-Dine baryogenesis just after inflation *Phys. Rev.* D **93** 083516

[17] Takahashi F and Yamada M 2016 Spontaneous baryogenesis from asymmetric inflation *Phys. Lett.* B **756** 216

[18] Dev P S B and Mohapatra R N 2015 TeV scale model for baryon and lepton number violation and resonant baryogenesis *Phys. Rev.* D **92** 016007

[19] Araki T *et al* 2005 Measurement of neutrino oscillation with KamLAND: evidence of spectral distortion *Phys. Rev. Lett.* **94** 8

[20] Yanagida T 1980 Horizontal symmetry and masses of neutrinos *Prog. Theor. Phys.* **64** 3
[21] Hamaguchi K 2002 Cosmological baryon asymmetry and neutrinos: baryogenesis via leptogenesis in supersymmetric theories arXiv: hep-ph-0212305
[22] Fukugita M and Yanagida T 1986 Baryogenesis without grand unification *Phys. Lett.* B **174** 1
[23] Davidson S, Nardi E and Nir Y 2008 Leptogenesis *Phys. Rep.* **466** 4
[24] Affleck I and Dine M 1985 A new mechanism for baryogenesis *Nucl. Phys.* B **249** 2
[25] Allahverdi R and Mazumdar A 2012 A mini review on Affleck–Dine baryogenesis *New J. Phys.* **14** 12
[26] Alvarez-Gaume L and Witten E 1984 Gravitational anomalies *Nucl. Phys.* B **234** 2
[27] Alexander S H S, Peskin M E and Sheikh-Jabbari M M 2006 Leptogenesis from gravity waves in models of inflation *Phys. Rev. Lett.* **96** 8
[28] Enqvist K and Anupam M 2003 Cosmological consequences of MSSM flat directions *Phys. Rep.* **380** 99–234

Chapter 12

Discussion and outlook

Electroweak baryogenesis is a rich framework which attempts to partially answer one of the most fundamental questions of our existence—why is there something rather than nothing? The answer it suggests is elegant and beautiful, but its most attractive features are that it requires new physics and it is testable. The next generation of colliders and EDM experiments should either confirm the necessary ingredients of EDMs or rule out the minimal scenario of the baryon asymmetry being produced during the transition to the Standard Model phase. Going beyond the minimal model via a multi-step electroweak phase transition [1–3] makes it possible to expand the parameter space where one expects new physics if electroweak baryogenesis is a correct history of our Universe. However, even in the multi-step scenario it remains to be seen if one can have the scale of new physics higher than what would be seen at the proposed 100 TeV collider without the scenario being contrived.

Even if nature decided that the matter–anti-matter asymmetry was indeed produced via the electroweak mechanism in a multi-step phase transition that puts the new physics scale out of our reach, there is another avenue through which EWBG could be tested. The electroweak mechanism requires that there be a strongly first-order phase transition in our cosmic history and such a phase transition leaves a relic gravitational wave background. If a phase transition occurred at either $T \in [10^7, 10^8]$ GeV, then the produced gravitational waves could be within the sensitivity ranges of LIGO, and further in the future LISA could verify a strongly first-order electroweak phase transition.

One theoretical issue we have glossed over in this analysis is the approximation that the mass basis and the vacuum expectation value vary slowly across a phase transition. The effect of the space–time varying mass matrix can be modeled through adding space–time-dependent transformation matrices, $U(x)$ to the Lagrangian [4, 5]. When we derive the equation of motion operators they now have a dependence on these transformation matrices. We proceed in the same way as before, acting on the Schwinger–Dyson equations from the left and right with the modified (that is

doi:10.1088/978-1-6817-4457-5ch12

transformation-matrix-dependent) equation of motion operators from the left and right. The sum of these two equations gives a constraint equation (which to lowest order in a suitable expansion will just give a space–time-dependent version of the usual dispersion relations) and the difference gives a kinetic equation. However, rather than taking the limit of $z \rightarrow x$, one rewrites the resulting constraint and kinetic equations in terms of the relative and center of mass coordinates—$x \mp z$, respectively—and performs the Wigner transform on the equations.

Solving the resulting integro-differential equations is a formidable task, but generic features can be teased out, in particular the important question as to whether there indeed is a resonance. Recent work involving toy models using the above described Wigner functional approach seems to indicate that there is resonance, with the caveat that the center of the resonance may be damped. Current results seem to suggest that there is enough of a resonance behavior left over to make many models of EWBG phenomenologically viable.

References

[1] Ovanesyan S I G and Ramsey-Musolf M J 2016 Two-step electroweak baryogenesis *Phys. Rev.* D **93** 015013

[2] Patel H H, Ramsey-Musolf M J and Wise M B 2013 Color breaking in the early universe *Phys. Rev.* D **88** 1

[3] Ramsey-Musolf M J, White G A and Winslow P 2016 Colour breaking baryogenesis in preparation

[4] Cirigliano V *et al* 2010 Flavored quantum Boltzmann equations *Phys. Rev.* D **81** 10

[5] Cirigliano V, Lee C and Tulin S 2011 Resonant flavor oscillations in electroweak baryogenesis *Phys. Rev.* D **84** 5

www.ingramcontent.com/pod-product-compliance
Lightning Source LLC
Chambersburg PA
CBHW082035230326
41598CB00081B/6517
* 9 7 8 1 6 8 1 7 4 4 5 6 8 *